PRIMARY MATHEMATICS 3B
TEXTBOOK

 **Marshall Cavendish**
Education

 SingaporeMath.com Inc

Original edition published under the title Primary Mathematics Textbook 3B

© 1982 Curriculum Planning & Development Division

Ministry of Education, Singapore

Published by Times Media Private Limited

This American Edition

© 2003 Times Media Private Limited

© 2003 Marshall Cavendish International (Singapore) Private Limited

Published by Marshall Cavendish Education

An imprint of Marshall Cavendish International (Singapore) Private Limited

Times Centre, 1 New Industrial Road, Singapore 536196

Customer Service Hotline: (65) 6411 0820

E-mail: fps@sg.marshallcavendish.com

Website: www.marshallcavendish.com/education

Distributed by

SingaporeMath.com Inc

404 Beavercreek Road #225

Oregon City, OR 97045

U.S.A.

Website: http://www.singaporemath.com

First published 2003

Second impression 2003

Reprinted 2004 (twice)

Third impression 2005

Reprinted 2006 (twice), 2007, 2008, 2009 (twice), 2010

ISBN 978-981-01-8503-9

Printed in Singapore by Times Printers, www.timesprinters.com

ACKNOWLEDGEMENTS

Our special thanks to Richard Askey, Professor of Mathematics (University of Wisconsin, Madison), Yoram Sagher, Professor of Mathematics (University of Illinois, Chicago), and Madge Goldman, President (Gabriella and Paul Rosenbaum Foundation), for their indispensable advice and suggestions in the production of Primary Mathematics (U.S. Edition).

PREFACE

Primary Mathematics (U.S. Edition) comprises textbooks and workbooks. The main feature of this package is the use of the **Concrete** ➡ **Pictorial** ➡ **Abstract** approach. The students are provided with the necessary learning experiences beginning with the concrete and pictorial stages, followed by the abstract stage to enable them to learn mathematics meaningfully. This package encourages active thinking processes, communication of mathematical ideas and problem solving.

The textbook comprises 6 units. Each unit is divided into parts: ❶, ❷, . . . Each part starts with a meaningful situation for communication and is followed by specific learning tasks numbered 1, 2, . . . The textbook is accompanied by a workbook. The sign Workbook Exercise> is used to link the textbook to the workbook exercises.

Practice exercises are designed to provide the students with further practice after they have done the relevant workbook exercises. Review exercises are provided for cumulative reviews of concepts and skills. All the practice exercises and review exercises are optional exercises.

The color patch ■ is used to invite active participation from the students and to facilitate oral discussion. The students are advised not to write on the color patches.

Challenging word problems are marked with *. Teachers may encourage the abler students to attempt them.

CONTENTS

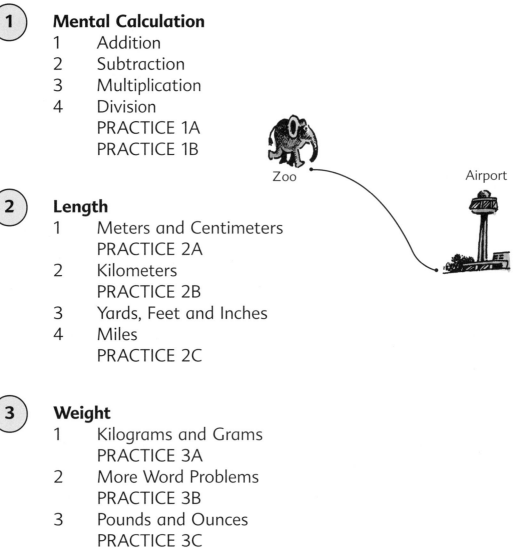

Zoo   Airport

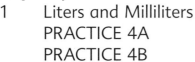

Mental Calculation

1 Addition

Add 46 and 27.

46 $\xrightarrow{+20}$ 66 $\xrightarrow{+7}$ 73

46 + 27 = ■

6

1. (a) 23 $\xrightarrow{+10}$ ■ $\xrightarrow{+4}$ ■

   23 + 14 = ■

   (b) 54 $\xrightarrow{+30}$ ■ $\xrightarrow{+6}$ ■

   54 + 36 = ■

   (c) 38 $\xrightarrow{+40}$ ■ $\xrightarrow{+5}$ ■

   38 + 45 = ■

2. (a) What number is 3 more than 68?
   (b) What number is 20 more than 94?

3. Add.
   (a) 43 + 30   (b) 67 + 20   (c) 85 + 50
   (d) 72 + 5   (e) 33 + 7   (f) 64 + 8
   (g) 36 + 23   (h) 27 + 35   (i) 55 + 26

   Workbook Exercise 1

4. Find the sum of 58 and 16.

   58 + 16 = ■

   58 + 16
     ╱ ╲
    2   14

   58 + 2 = 60
   58 + 16 = 60 + 14

5. Add.
   (a) 39 + 27   (b) 58 + 34   (c) 45 + 65

   Workbook Exercise 2

2 Subtraction

Subtract 34 from 87.

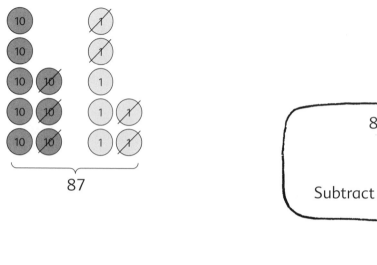

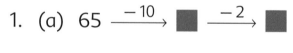

87

$$87 \xrightarrow{-30} 57 \xrightarrow{-4} 53$$

$$87 - 34 = \blacksquare$$

1. (a) $65 \xrightarrow{-10} \blacksquare \xrightarrow{-2} \blacksquare$

   $65 - 12 = \blacksquare$

   (b) $76 \xrightarrow{-40} \blacksquare \xrightarrow{-6} \blacksquare$

   $76 - 46 = \blacksquare$

   (c) $63 \xrightarrow{-20} \blacksquare \xrightarrow{-8} \blacksquare$

   $63 - 28 = \blacksquare$

2. (a) What number is 2 less than 51?
    (b) What number is 30 less than 76?

3. Subtract.
    (a) 70 − 30   (b) 95 − 70
    (c) 68 − 60   (d) 58 − 6
    (e) 83 − 3   (f) 47 − 9

4. Subtract.
    (a) 48 − 32   (b) 64 − 34
    (c) 85 − 59   (d) 56 − 24
    (e) 87 − 47   (f) 63 − 55

5. Find the difference between 90 and 18.

    $90 - 18 =$ ■

    $$90 - 18$$
    70   20

    $20 - 18 = 2$
    $90 - 18 = 70 + 2$

6. Subtract.
    (a) 30 − 28   (b) 60 − 56
    (c) 70 − 65   (d) 50 − 17
    (e) 40 − 29   (f) 80 − 58
    (g) 40 − 16   (h) 70 − 58
    (i) 90 − 39

Workbook Exercise 3

③ Multiplication

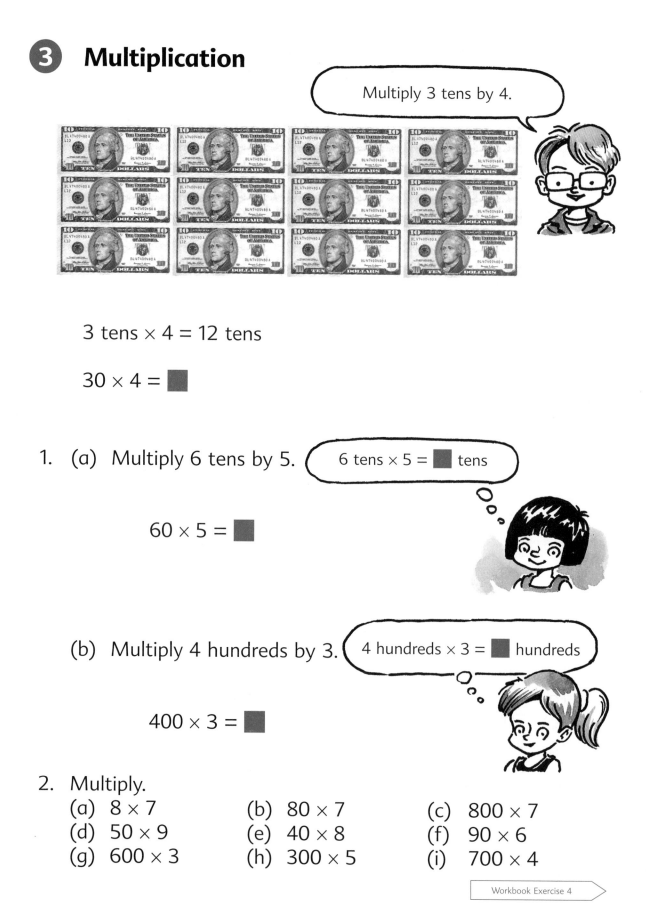

Multiply 3 tens by 4.

3 tens × 4 = 12 tens

30 × 4 = ■

1. (a) Multiply 6 tens by 5.

6 tens × 5 = ■ tens

60 × 5 = ■

(b) Multiply 4 hundreds by 3.

4 hundreds × 3 = ■ hundreds

400 × 3 = ■

2. Multiply.
   (a) 8 × 7   (b) 80 × 7   (c) 800 × 7
   (d) 50 × 9   (e) 40 × 8   (f) 90 × 6
   (g) 600 × 3   (h) 300 × 5   (i) 700 × 4

Workbook Exercise 4

4 Division

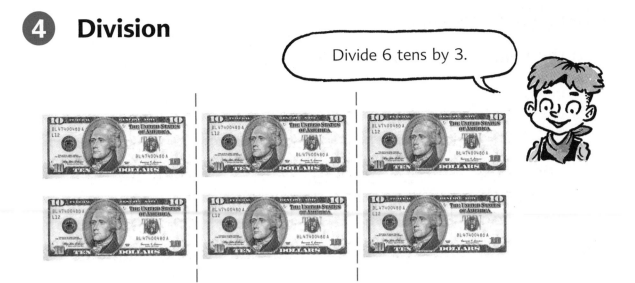

Divide 6 tens by 3.

6 tens ÷ 3 = 2 tens

$60 \div 3 =$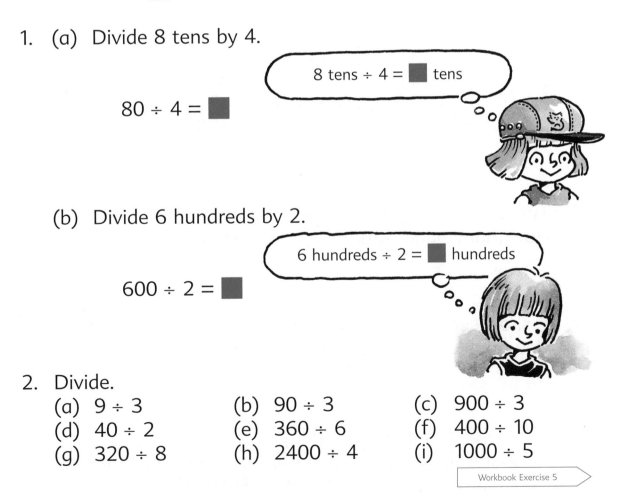

1. (a) Divide 8 tens by 4.

8 tens ÷ 4 = ▮ tens

$80 \div 4 =$ ▮

(b) Divide 6 hundreds by 2.

6 hundreds ÷ 2 = ▮ hundreds

$600 \div 2 =$ ▮

2. Divide.
   (a) $9 \div 3$
   (b) $90 \div 3$
   (c) $900 \div 3$
   (d) $40 \div 2$
   (e) $360 \div 6$
   (f) $400 \div 10$
   (g) $320 \div 8$
   (h) $2400 \div 4$
   (i) $1000 \div 5$

Workbook Exercise 5

PRACTICE 1A

Find the value of each of the following:

	(a)	(b)	(c)
1.	65 + 28	34 + 66	18 + 84
2.	99 + 99	99 + 98	27 + 45
3.	78 − 45	90 − 56	90 − 85
4.	99 − 98	83 − 75	98 − 97
5.	4 × 30	50 × 8	200 × 9
6.	500 ÷ 5	600 ÷ 10	400 ÷ 2
7.	100 × 6	5 × 90	300 × 7
8.	160 ÷ 8	240 ÷ 3	810 ÷ 9
9.	6 × 100	40 × 6	500 × 5

10. (a) What number is 29 less than 84?
    (b) What number is 68 less than 310?
    (c) What number is 35 more than 475?
    (d) What number is 97 more than 5397?

11. There are 80 pages in one exercise book.
    How many pages are there in 6 exercise books?

12. A shopkeeper packed 200 onions equally into 5 bags.
    How many onions were there in each bag?

13. Alex sold 70 muffins on Friday.
    He sold 4 times as many muffins on Sunday as on Friday.
    How many muffins did he sell on Sunday?

PRACTICE 1B

Find the value of each of the following:

	(a)	(b)	(c)
1.	50×3	60×5	30×6
2.	$40 \div 2$	$80 \div 4$	$600 \div 3$
3.	6×90	7×400	300×8
4.	$270 \div 9$	$320 \div 8$	$720 \div 9$

5. Alice saves $50 a month.
   How much does she save in 8 months?

6. A cake shop sold 200 muffins.
   It sold 4 times as many muffins as chocolate cakes.
   How many chocolate cakes did it sell?

7. A fruit seller bought 9 boxes of pears.
   There were 40 pears in each box.
   How many pears did he buy altogether?

8. There are 6 coins in a set.
   How many coins are there in 200 sets?

9. Chris packed 250 tomatoes into bags of 5 each.
   (a) How many bags of tomatoes were there?
   (b) He sold all the tomatoes at $2 a bag.
       How much money did he receive?

10. (a) Brian bought 98 blue pens and 62 red pens. How many
        pens did he buy altogether?
    (b) He divided the pens equally into 8 boxes. How many
        pens were there in each box?

2 Length

1 Meters and Centimeters

Get a meter ruler and find out how long 1 meter is.

Estimate the length of the chalkboard in the classroom.
Then check by measuring it with the meter ruler.

Is the length closer to 2 m or 3 m?

Estimate the height of the door in your classroom.
Check by measuring.

Is the door 2 m high?

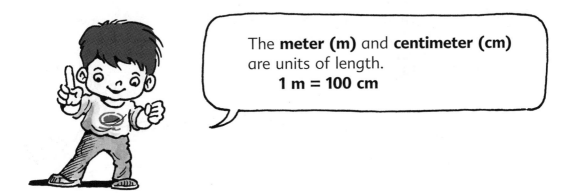

The **meter (m)** and **centimeter (cm)** are units of length.
1 m = 100 cm

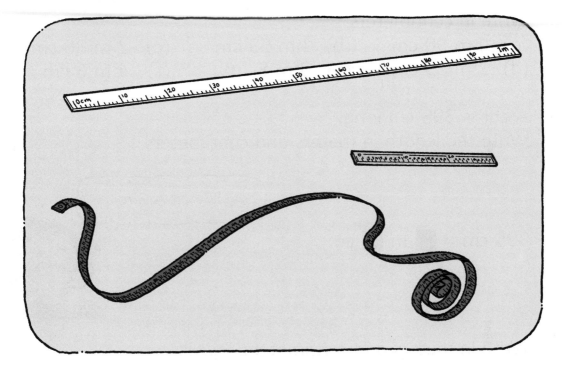

1. Joe's height is 1 m 25 cm.
   (a) 1 m 25 cm is ■ cm more than 1 m.
   (b) 1 m 25 cm = ■ cm

2. (a) Write 2 m in centimeters.
       2 m = ■ cm
   (b) Write 300 cm in meters.
       300 cm = ■ m

3. Walk 5 steps.
   Measure the distance in meters and centimeters.
   Then write the distance in centimeters.

15

4. David's long jump result is 1 m 45 cm.
   Write the distance in centimeters.

   1 m 45 cm =  cm

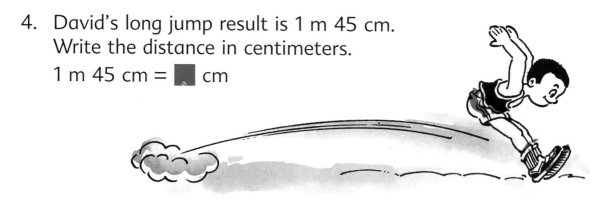

5. Write in centimeters.
   (a) 1 m 90 cm   (b) 1 m 55 cm   (c) 2 m 86 cm
   (d) 2 m 89 cm   (e) 3 m 8 cm   (f) 4 m 6 cm

6. A car is 395 cm long.
   Write the length in meters and centimeters.

   395 cm = ■ m ■ cm

   300 cm = 3 m

7. Write in meters and centimeters.
   (a) 180 cm   (b) 195 cm   (c) 262 cm
   (d) 299 cm   (e) 304 cm   (f) 409 cm

8. The table shows the results of the shot put finals.

Name	Distance
Ryan	1 m 89 cm
Andy	2 m 08 cm
Tyrone	1 m 96 cm

   Arrange the distances in order. Begin with the shortest.

Workbook Exercise 6

9. Lily has a piece of red ribbon 3 m 40 cm long and a piece of yellow ribbon 1 m 85 cm long.

(a) Find the total length of the ribbons.

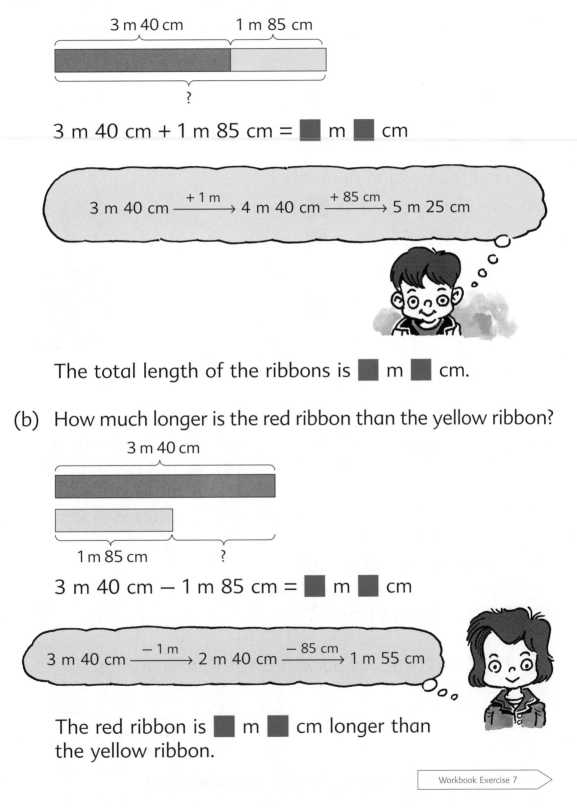

3 m 40 cm + 1 m 85 cm = ■ m ■ cm

$$3 \text{ m } 40 \text{ cm } \xrightarrow{+1 \text{ m}} 4 \text{ m } 40 \text{ cm } \xrightarrow{+85 \text{ cm}} 5 \text{ m } 25 \text{ cm}$$

The total length of the ribbons is ■ m ■ cm.

(b) How much longer is the red ribbon than the yellow ribbon?

3 m 40 cm − 1 m 85 cm = ■ m ■ cm

$$3 \text{ m } 40 \text{ cm } \xrightarrow{-1 \text{ m}} 2 \text{ m } 40 \text{ cm } \xrightarrow{-85 \text{ cm}} 1 \text{ m } 55 \text{ cm}$$

The red ribbon is ■ m ■ cm longer than the yellow ribbon.

Workbook Exercise 7

PRACTICE 2A

1. Write in centimeters.
   (a) 4 m
   (b) 1 m 40 cm
   (c) 2 m 25 cm
   (d) 3 m 95 cm
   (e) 4 m 5 cm
   (f) 9 m 9 cm

2. Write in meters and centimeters.
   (a) 120 cm
   (b) 225 cm
   (c) 309 cm
   (d) 618 cm
   (e) 963 cm
   (f) 405 cm

3. Find the missing numbers.

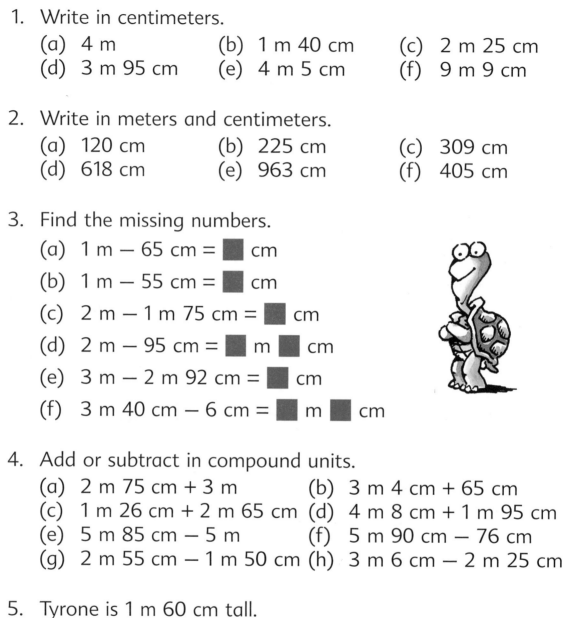

   (a) 1 m − 65 cm = ■ cm
   (b) 1 m − 55 cm = ■ cm
   (c) 2 m − 1 m 75 cm = ■ cm
   (d) 2 m − 95 cm = ■ m ■ cm
   (e) 3 m − 2 m 92 cm = ■ cm
   (f) 3 m 40 cm − 6 cm = ■ m ■ cm

4. Add or subtract in compound units.
   (a) 2 m 75 cm + 3 m
   (b) 3 m 4 cm + 65 cm
   (c) 1 m 26 cm + 2 m 65 cm
   (d) 4 m 8 cm + 1 m 95 cm
   (e) 5 m 85 cm − 5 m
   (f) 5 m 90 cm − 76 cm
   (g) 2 m 55 cm − 1 m 50 cm
   (h) 3 m 6 cm − 2 m 25 cm

5. Tyrone is 1 m 60 cm tall.
   Ryan is 16 cm shorter than Tyrone.
   What is Ryan's height?

6. Mr. Chen tied two packages with these strings.

   1 m 80 cm   1 m 65 cm

   What was the total length of the strings?

Kilometers

The **kilometer (km)**, meter (m) and centimeter (cm) are units of length.

1 km = 1000 m
1 m = 100 cm

A bus is about 10 m long,
The total length of 100 buses
is about 1 km.

We measure long distances
in kilometers.

ECP
1 km

1.

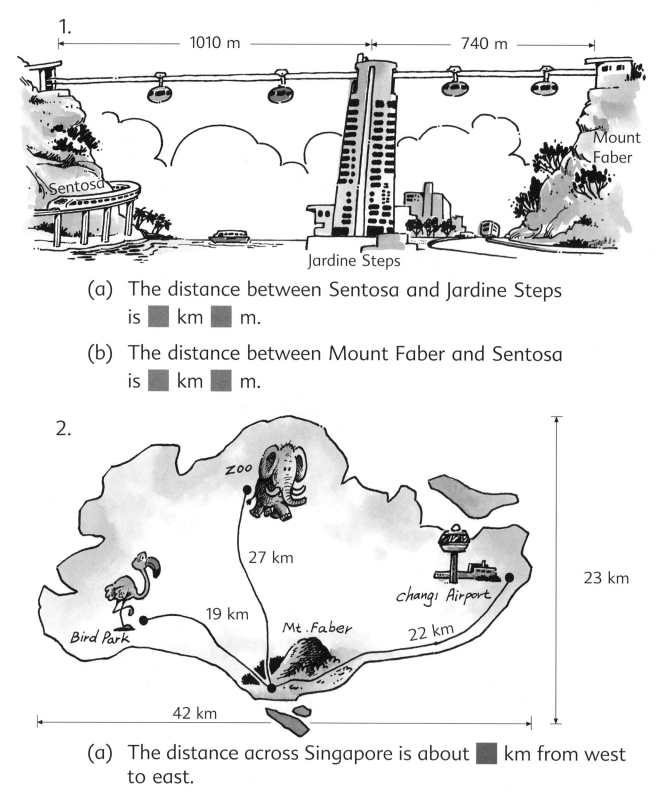

1010 m   740 m

Sentosa

Mount Faber

Jardine Steps

(a) The distance between Sentosa and Jardine Steps is ▮ km ▮ m.

(b) The distance between Mount Faber and Sentosa is ▮ km ▮ m.

2.

Zoo

27 km

19 km

Bird Park

Mt. Faber

changi Airport

22 km

23 km

42 km

(a) The distance across Singapore is about ▮ km from west to east.
It is about ▮ km from north to south.

(b) The distance from the Bird Park to Changi Airport is about ▮ km.

3.

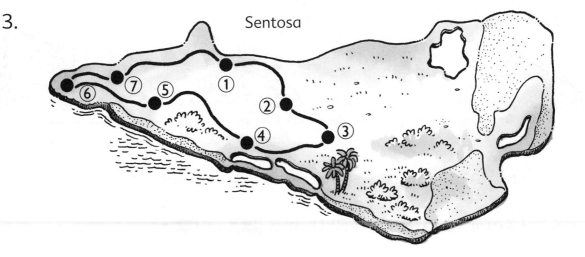

The total length of the Monorail route on Sentosa Island is about 6 km 100 m.
Write the length in meters.

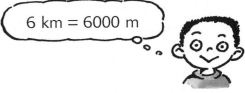

6 km = 6000 m

4. The distance around a running track is 400 m.
Ryan ran round the track 3 times.

He ran ▮ km ▮ m.

5. Write in meters.
   (a) 1 km 600 m   (b) 2 km 550 m   (c) 2 km 605 m
   (d) 3 km 85 m   (e) 3 km 20 m   (f) 4 km 5 m

6. Write in kilometers and meters.
   (a) 1830 m   (b) 2304 m   (c) 2780 m
   (d) 3096 m   (e) 3040 m   (f) 4009 m

Workbook Exercise 8

7.

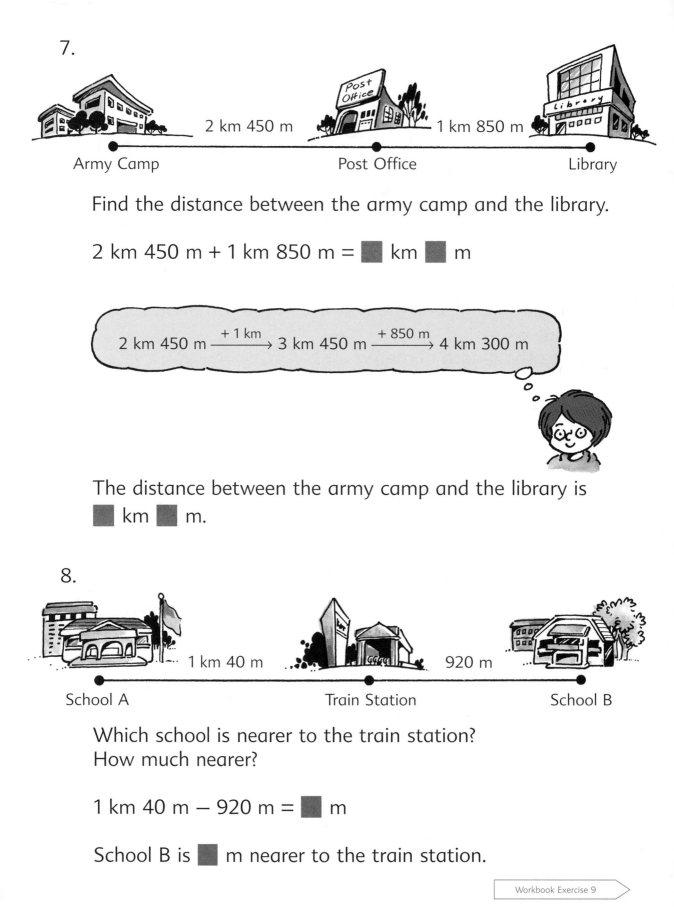

Army Camp   2 km 450 m   Post Office   1 km 850 m   Library

Find the distance between the army camp and the library.

2 km 450 m + 1 km 850 m = █ km █ m

2 km 450 m $\xrightarrow{+ 1 km}$ 3 km 450 m $\xrightarrow{+ 850 m}$ 4 km 300 m

The distance between the army camp and the library is █ km █ m.

8.

School A   1 km 40 m   Train Station   920 m   School B

Which school is nearer to the train station?
How much nearer?

1 km 40 m − 920 m = █ m

School B is █ m nearer to the train station.

Workbook Exercise 9

PRACTICE 2B

1. Write in meters.
   - (a) 3 km
   - (b) 1 km 450 m
   - (c) 2 km 506 m
   - (d) 2 km 60 m
   - (e) 3 km 78 m
   - (f) 4 km 9 m

2. Write in kilometers and meters.
   - (a) 1680 m
   - (b) 1085 m
   - (c) 2204 m
   - (d) 3090 m
   - (e) 3999 m
   - (f) 4001 m

3. Find the missing numbers.

   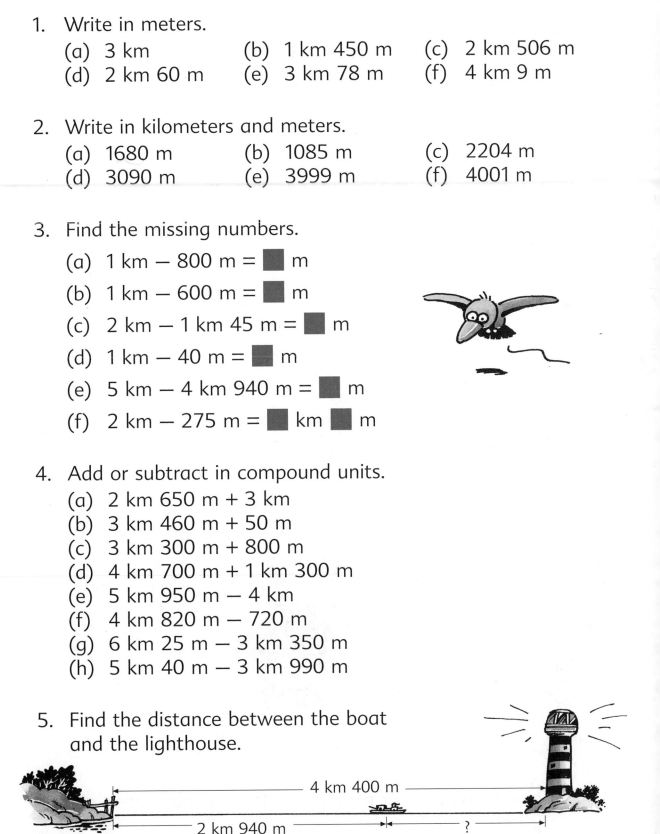

   - (a) 1 km — 800 m = ■ m
   - (b) 1 km — 600 m = ■ m
   - (c) 2 km — 1 km 45 m = ■ m
   - (d) 1 km — 40 m = ■ m
   - (e) 5 km — 4 km 940 m = ■ m
   - (f) 2 km — 275 m = ■ km ■ m

4. Add or subtract in compound units.
   - (a) 2 km 650 m + 3 km
   - (b) 3 km 460 m + 50 m
   - (c) 3 km 300 m + 800 m
   - (d) 4 km 700 m + 1 km 300 m
   - (e) 5 km 950 m — 4 km
   - (f) 4 km 820 m — 720 m
   - (g) 6 km 25 m — 3 km 350 m
   - (h) 5 km 40 m — 3 km 990 m

5. Find the distance between the boat and the lighthouse.

   4 km 400 m

   2 km 940 m   ?

23

3 Yards, Feet and Inches

Get a yard stick and find out how long 1 yard is.
Estimate the length of the teacher's desk in your classroom.
Check by measuring it.

Is the length of your teacher's desk more than 2 yards?

Get a ruler and find out how long 1 foot is.
Estimate the length of your math book.
Check by measuring with your ruler.

Is your math book less than a foot long?

Take a look at your ruler again. Is your math book about 10 inches long?

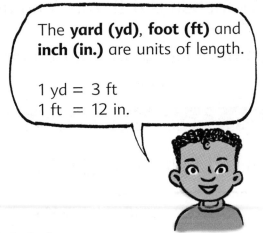

The **yard (yd)**, **foot (ft)** and **inch (in.)** are units of length.

1 yd = 3 ft
1 ft = 12 in.

1. A table is 1 yd 2 ft long.
   (a) 1 yd 2 ft is ▪ ft more than 1 yd.
   (b) 1 yd 2 ft = ▪ ft

2. (a) Write 8 yd in feet.
      8 yd = ▪ ft

   (b) Write 18 ft in yards.
      18 ft = ▪ yd

3. A piece of ribbon is 11 yd 2 ft.
   Write the length in feet.

4. The length of the sofa is 15 ft long.
   The sofa is ▪ yd long.

5. The coffee table is 1 ft 5 in. high.
   (a) 1 ft 5 in. is ▪ in. more than 1 ft.
   (b) 1 ft 5 in. = ▪ in.

6. (a) Write 7 ft in inches.
      7 ft = ▪ in.

7. Measure the width of your classroom door in feet and inches. Then write the width in inches.

> To change yards to feet, we multiply by 3.
>
> 1 yd = 1 × 3 = 3 ft
> 2 yd = 2 × 3 = 6 ft
> 3 yd = 3 × 3 = 9 ft

> To change feet to yards, we divide by 3.
>
> 3 ft = 3 ÷ 3 = 1 yd
> 6 ft = 6 ÷ 3 = 2 yd

> To change feet to inches, we multiply by 12.
>
> 1 ft = 1 × 12 = 12 in.
> 2 ft = 2 × 12 = 24 in.

Workbook Exercise 10

8. A piece of blue ribbon is 4 ft 7 in. long and a piece of green ribbon is 1 ft 10 in. long.

(a) Find the total length of the ribbons.

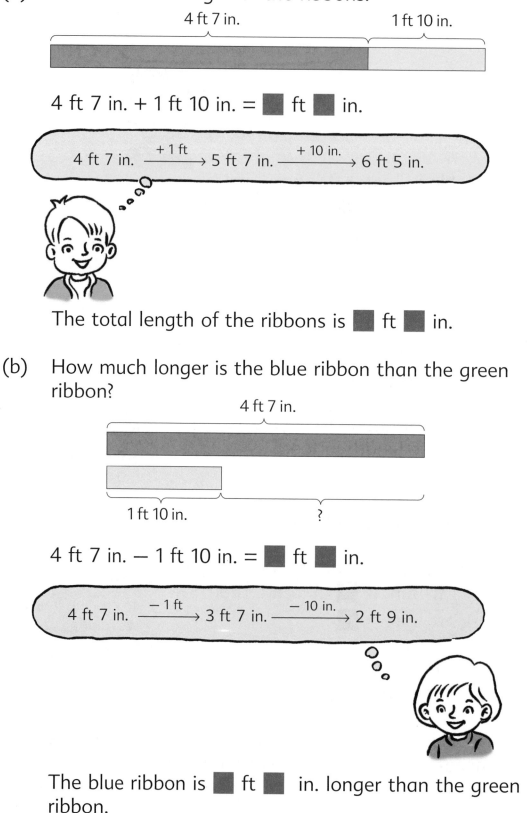

4 ft 7 in. + 1 ft 10 in. = ⬛ ft ⬛ in.

4 ft 7 in. $\xrightarrow{\;+\,1\,ft\;}$ 5 ft 7 in. $\xrightarrow{\;+\,10\,in.\;}$ 6 ft 5 in.

The total length of the ribbons is ⬛ ft ⬛ in.

(b) How much longer is the blue ribbon than the green ribbon?

4 ft 7 in.

1 ft 10 in.   ?

4 ft 7 in. − 1 ft 10 in. = ⬛ ft ⬛ in.

4 ft 7 in. $\xrightarrow{\;-\,1\,ft\;}$ 3 ft 7 in. $\xrightarrow{\;-\,10\,in.\;}$ 2 ft 9 in.

The blue ribbon is ⬛ ft ⬛ in. longer than the green ribbon.

4 Miles

The **mile (mi)**, yard (yd), foot (ft) and inch (in.) are units of length.

1 mile = 5280 feet

We measure long distances in miles.

1. Juan jogged 3 mi daily. How many miles did he jog in a week?

2.
San Francisco   Denver   New York

◄——— 1260 mi ———►

◄——————— 2930 mi ———————►

The distance between San Francisco and New York is ▮ mi.

Find the distance between New York and Denver.

Workbook Exercise 11

PRACTICE 2C

1. Write in feet.
    (a) 5 yd   (b) 87 yd 2 ft   (c) 308 yd 1 ft

2. Write in inches.
    (a) 9 ft   (b) 6 ft 10 in.   (c) 9 ft 9 in.

3. Write in yards and feet.
    (a) 27 ft   (b) 108 ft   (c) 212 ft

4. Write in feet and inches.
    (a) 12 in.   (b) 16 in.   (c) 24 in.

5. Find the missing numbers.
    (a) 1 yd − 2 ft = ■ ft
    (b) 2 yd − 1 yd 1 ft = ■ ft
    (c) 15 yd − 14 yd 2 ft = ■ ft
    (d) 1 ft − 7 in. = ■ in.
    (e) 13 ft − 12 ft 8 in. = ■ in.

6. Add or subtract in compound units.
    (a) 5 yd 1 ft + 1 ft
    (b) 8 yd 2 ft + 2 ft
    (c) 5 yd 1 ft + 2 yd 2 ft
    (d) 3 yd 2 ft − 2 yd
    (e) 9 yd 1 ft − 2 ft
    (f) 6 yd 1 ft − 5 yd 2 ft

7. Add or subtract in compound units.
    (a) 5 ft 11 in. + 7 ft
    (b) 9 ft 5 in. + 6 in.
    (c) 8 ft 7 in. + 3 ft 8 in.
    (d) 11 ft 11 in. − 6 in.
    (e) 10 ft 4 in. − 6 in.
    (f) 6 ft 8 in. − 4 ft 10 in.

Weight

1 Kilograms and Grams

The **kilogram (kg)** and **gram (g)** are units of weight.

1 kg = 1000 g

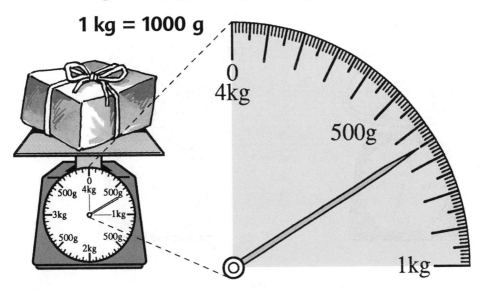

The package weighs 650 g.

The grapes weigh
⬛ g.

The papaya weighs
⬛ kg ⬛ g.

1. Read the scales.

(a)

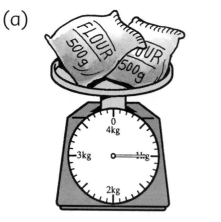

(b)

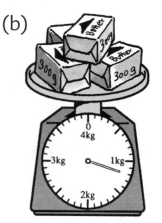

(c)

(d)

2. The potatoes weigh 2 kg 200 g.
   Write the weight in grams.

2 kg = 2000 g

3. Each book weighs 350 g.
   The total weight of 4 books
   is ▉ kg ▉ g.

Workbook Exercise 12

4. A bag of peanuts weighs 1 kg 850 g.
   How much more peanuts are needed to make up 2 kg?

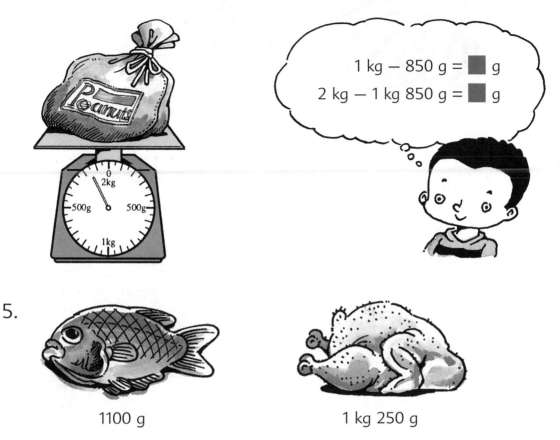

1 kg − 850 g = ▮ g

2 kg − 1 kg 850 g = ▮ g

5.

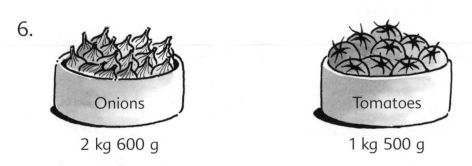

1100 g   1 kg 250 g

Which is heavier, the fish or the chicken?
How much heavier?

6.

Onions   Tomatoes

2 kg 600 g   1 kg 500 g

(a) What is the total weight of the onions and the
    tomatoes?
(b) What is the difference in weight between the onions
    and the tomatoes?

Workbook Exercise 13

7.

3 kg 80 g   1 kg 960 g

(a) Find the total weight of the watermelon and the bananas.

3 kg 80 g + 1 kg 960 g = ■ kg ■ g

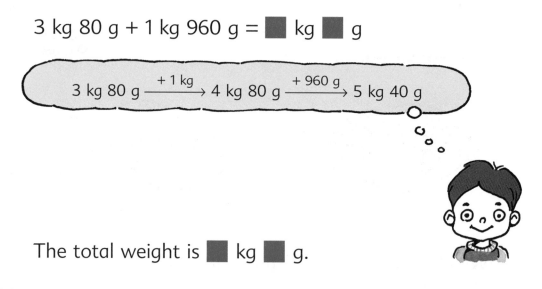

3 kg 80 g $\xrightarrow{+1 \text{ kg}}$ 4 kg 80 g $\xrightarrow{+960 \text{ g}}$ 5 kg 40 g

The total weight is ■ kg ■ g.

(b) Find the difference in weight between the watermelon and the bananas.

3 kg 80 g − 1 kg 960 g = ■ kg ■ g

3 kg 80 g $\xrightarrow{-1 \text{ kg}}$ 2 kg 80 g $\xrightarrow{-960 \text{ g}}$ 1 kg 120 g

The difference in weight is ■ kg ■ g.

Workbook Exercise 14

PRACTICE 3A

1. Write in grams.
   (a) 1 kg 456 g   (b) 2 kg 370 g   (c) 3 kg 808 g
   (d) 2 kg 80 g   (e) 1 kg 8 g   (f) 4 kg 7 g

2. Write in kilograms and grams.
   (a) 2143 g   (b) 1354 g   (c) 3800 g
   (d) 2206 g   (e) 3085 g   (f) 4009 g

3. Find the missing numbers.
   (a) 1 kg − 395 g = ▇ g
   (b) 1 kg − 85 g = ▇ g
   (c) 3 kg − 2 kg 400 g = ▇ g
   (d) 5 kg − 4 kg 60 g = ▇ g
   (e) 1 kg − 540 g = ▇ g
   (f) 3 kg − 805 g = ▇ kg ▇ g

4. Add or subtract in compound units.
   (a) 3 kg 500 g + 2 kg   (b) 4 kg 650 g + 450 g
   (c) 3 kg 100 g + 1 kg 900 g   (d) 2 kg 50 g + 4 kg 70 g
   (e) 3 kg 10 g − 200 g   (f) 4 kg 300 g − 1 kg 50 g
   (g) 4 kg 250 g − 1 kg 500 g   (h) 5 kg − 2 kg 905 g

5. Lily weighed 25 kg 750 g two years ago.
   Now she weighs 32 kg.
   How much weight did she gain?

6. A pumpkin weighs 2 kg 990 g.
   A watermelon weighs 4 kg 200 g.
   (a) Find the total weight of the pumpkin and the
       watermelon.
   (b) Find the difference in weight between the pumpkin and
       the watermelon.

33

2 More Word Problems

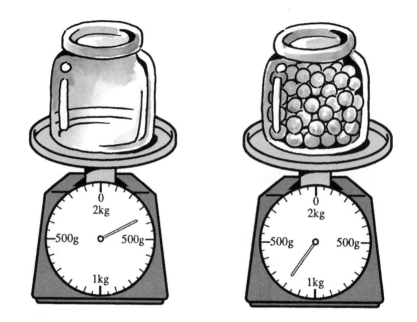

Weight of empty jar	+	Weight of marbles	=	Total weight of jar and marbles
(350 g)		(?)		(1 kg 200 g)

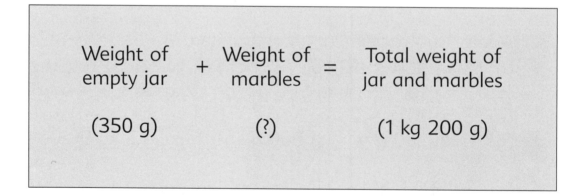

Weight of marbles = 1 kg 200 g − 350 g

= ■ g

1. A bottle of sauce weighs 560 g.
    The empty bottle weighs 305 g.
    How many grams of sauce are there in the bottle?

2. A basket of fruits weighs 1 kg 60 g.
    The empty basket weighs 200 g.
    Find the weight of the fruits.

3. William's weight is 57 kg.
    He is 3 times as heavy as Sean.
    What is Sean's weight?

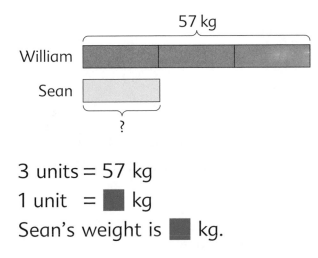

    3 units = 57 kg
    1 unit = ■ kg
    Sean's weight is ■ kg.

4. A watermelon is 5 times as heavy as a papaya.
    If the papaya weighs 950 g, find the weight of the
    watermelon.

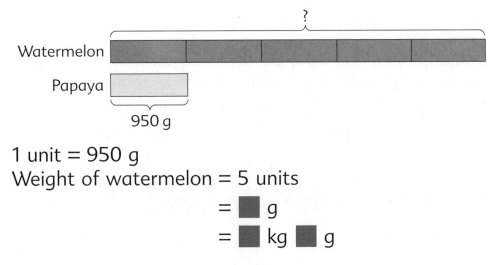

    1 unit = 950 g
    Weight of watermelon = 5 units
                         = ■ g
                         = ■ kg ■ g

5. John weighs 34 kg 600 g.
   He is 800 g heavier than David.
   What is David's weight?

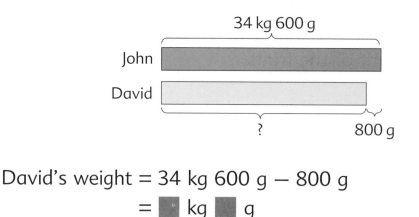

   David's weight = 34 kg 600 g − 800 g
   $\qquad\qquad$ = ⬛ kg ⬛ g

6. A goose weighs 3 kg 200 g.
   A duck weighs 1 kg 800 g.
   (a) What is the total weight of the goose and the duck?
   (b) What is the difference in weight between the goose
       and the duck?

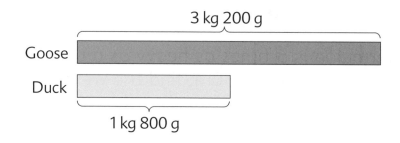

7. A pineapple weighs 2 kg 50 g.
   A watermelon is 600 g heavier than the pineapple.
   (a) What is the weight of the watermelon?
   (b) What is the total weight of the two fruits?

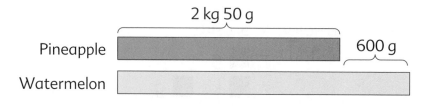

8. The total weight of a bag of flour and a bag of salt is 2 kg 400 g.
   If the bag of flour weighs 1 kg 950 g, find the weight of the bag of salt.

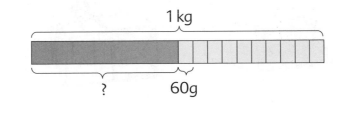

2 kg 400 g

1 kg 950 g   ?

9. The total weight of a football and 10 tennis balls is 1 kg.
   If the weight of each tennis ball is 60 g, find the weight of the football.

1 kg

?   60g

Weight of 10 tennis balls = 60 × 10
                          = 600 g

Weight of the football   = 1 kg − 600 g
                          = ▮ g

10. The total weight of a bottle of cooking oil and 2 bags of sugar is 5 kg 50 g.
    If the weight of each bag of sugar is 2 kg, find the weight of the bottle of cooking oil.

Weight of 2 bags of sugar   = ▮ kg
Weight of the bottle of cooking oil = ▮ kg ▮ g

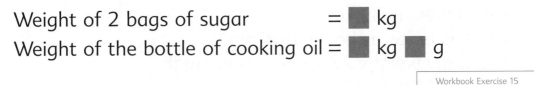

Workbook Exercise 15

PRACTICE 3B

1. Write in grams.
    (a) 5 kg   (b) 1 kg 950 g (c) 1 kg 60 g
    (d) 2 kg 805 g (e) 2 kg 5 g   (f) 3 kg 2 g

2. Write in kilograms and grams.
    (a) 1905 g   (b) 1055 g   (c) 2208 g
    (d) 3390 g   (e) 3599 g   (f) 5002 g

3. Add or subtract in compound units.
    (a) 2 kg 940 g + 300 g (b) 3 kg 880 g + 1 kg 220 g
    (c) 4 kg − 1 kg 480 g (d) 5 kg 20 g − 2 kg 450 g

4.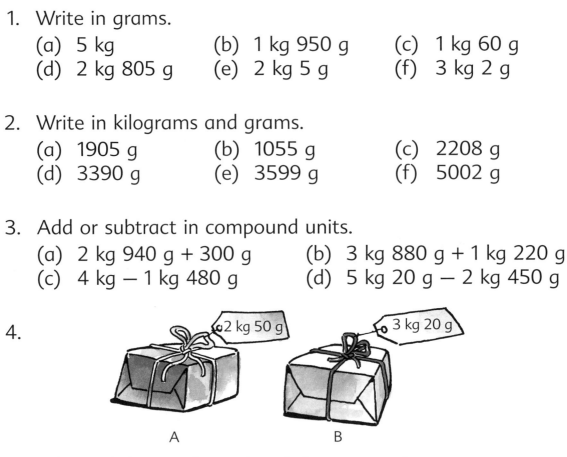

    (a) Find the total weight of the two packages.
    (b) Find the difference in weight between the
         two packages.

5. The total weight of Ali and Sam is 100 kg.
    If Ali's weight is 46 kg 540 g, find Sam's weight.

6. Brian's weight is 70 kg.
    He is 5 times as heavy as his son.
    Find the total weight of Brian and his son.

7. David weighs 39 kg.
    Hugh is twice as heavy as David.
    Matthew weighs 27 kg less than Hugh.
    What is Matthew's weight?

③ Pounds and Ounces

The **pound (lb)** and **ounce (oz)** are units of weight.

1 lb = 16 oz

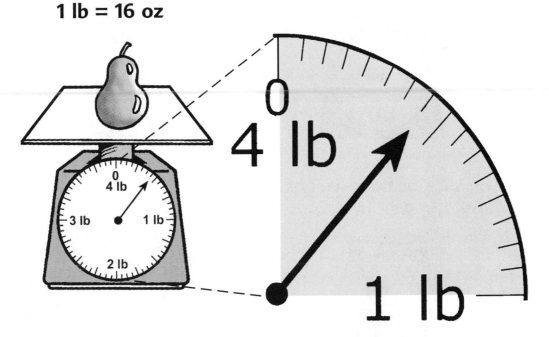

The pear weighs 7 oz.

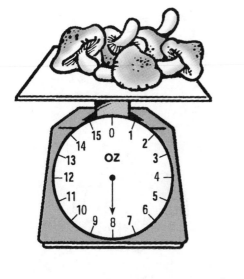

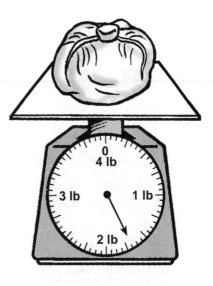

The mushrooms weigh
⬛ oz.

The cabbage weighs
⬛ lb ⬛ oz.

1. Read the scales.

(a)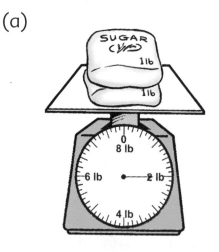

(b)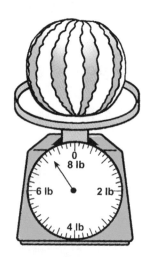

2. The sack of flour weighs 4 lb 13 oz.
   Write the weight in oz.

1 lb = 1 × 16 = 16 oz
4 lb = 4 × 16 = 64 oz

3. Each stick of butter weighs 8 oz.
   The total weight of 3 sticks of
   butter is ▮ lb ▮ oz.

4.

6 lb 15 oz   115 oz

Which is lighter, the potatoes or the pumpkin?

5.

1 lb 14 oz   3 lb 9 oz

(a) Find the total weight of the cantaloupe and the grapes.

3 lb 9 oz + 1 lb 14 oz = ▮ lb ▮ oz

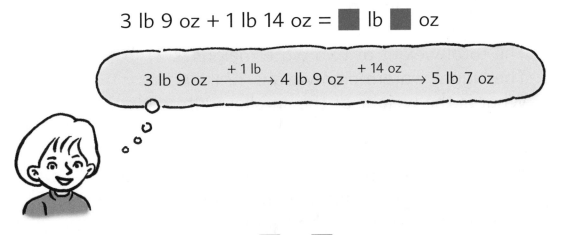

$$3 \text{ lb } 9 \text{ oz} \xrightarrow{+ 1 \text{ lb}} 4 \text{ lb } 9 \text{ oz} \xrightarrow{+ 14 \text{ oz}} 5 \text{ lb } 7 \text{ oz}$$

The total weight is ▮ lb ▮ oz.

(b) Find the difference in weight between the cantaloupe and the grapes.

3 lb 9 oz − 1 lb 14 oz = ▮ lb ▮ oz

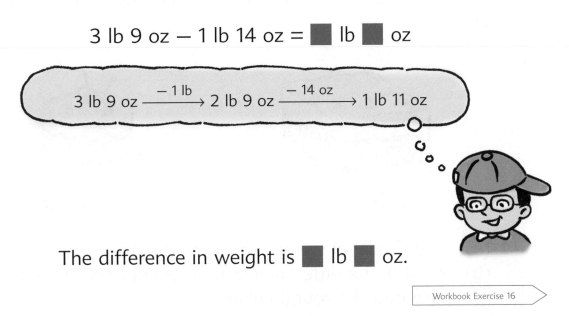

$$3 \text{ lb } 9 \text{ oz} \xrightarrow{- 1 \text{ lb}} 2 \text{ lb } 9 \text{ oz} \xrightarrow{- 14 \text{ oz}} 1 \text{ lb } 11 \text{ oz}$$

The difference in weight is ▮ lb ▮ oz.

Workbook Exercise 16

PRACTICE 3C

1. Write in ounces.
   (a) 5 lb   (b) 7 lb 15 oz   (c) 9 lb 9 oz

2. Write in pounds and ounces.
   (a) 16 oz   (b) 20 oz   (c) 26 oz

3. Add or subtract in compound units.
   (a) 7 lb 5 oz + 14 oz   (b) 5 lb 11 oz + 3 lb 11 oz
   (c) 12 lb − 8 lb 2 oz   (d) 3 lb 1 oz − 1 lb 15 oz

4. The total weight of two watermelons is 21 lb.
   The bigger watermelon weighs 12 lb 9 oz.
   What is the weight of the smaller watermelon?

5. A tomato weighs 3 oz.
   An avocado weighs 4 oz more than the tomato.
   A squash weighs twice as heavy as the avocado.
   What is the weight of the squash?

6. A pumpkin weighs 21 lb.
   The pumpkin is 7 times as heavy as a bunch of bananas.
   What is the total weight of the pumpkin and the bunch of bananas?

7.

3 lb 7 oz   2 lb 10 oz

   (a) Find the total weight of the two cakes.
   (b) What is the difference in weight between the square cake and the round cake?

REVIEW A

Find the value of each of the following:

	(a)	(b)	(c)
1.	499 + 42	507 + 3593	3084 + 63
2.	750 − 145	1806 − 82	7009 − 5
3.	53 × 7	156 × 5	407 × 4
4.	87 ÷ 3	104 ÷ 8	324 ÷ 6

5. There are 24 boxes of chocolates in a carton.
   How many boxes of chocolates are there in 8 cartons?

6. 5 people shared $450 equally.
   How much money did each person receive?

7. Lindsey gave each of her friends 7 cupcakes.
   She gave away 140 cupcakes altogether.
   How many friends did she give the cupcakes to?

8. There are 8 boxes of yellow and green buttons.
   There are 46 buttons in each box.
   If there are 200 yellow buttons, how many green buttons
   are there?

9. There were 150 bulbs in a box.
   6 of them were broken.
   The rest were packed into boxes of 4 bulbs each.
   How many boxes of bulbs were there?

10. Mr. Wang bought 3 boxes of oranges.
    There were 36 oranges in the first box.
    There were 54 oranges in each of the other two boxes.
    How many oranges did he buy?

43

REVIEW B

Find the missing numbers.

1. (a) 5 m = ▮ cm   (b) 4 m 8 cm = ▮ cm
   (c) 2 km 560 m = ▮ m   (d) 3 km 5 m = ▮ m
   (e) 1 kg 30 g = ▮ g   (f) 2 kg 80 g = ▮ g

2. (a) 208 cm = ▮ m ▮ cm
   (b) 320 cm = ▮ m ▮ cm
   (c) 1850 m = ▮ km ▮ m
   (d) 2004 m = ▮ km ▮ m
   (e) 3095 g = ▮ kg ▮ g
   (f) 4209 g = ▮ kg ▮ g

Add or subtract in compound units.

3. (a) 1 m 58 cm + 70 cm   (b) 2 m 95 cm + 2 m 45 cm
   (c) 3 m − 2 m 35 cm   (d) 4 m 5 cm − 1 m 85 cm

4. (a) 5 km 690 m + 520 m   (b) 7 km 960 m + 2 km 240 m
   (c) 9 km 420 m − 780 m   (d) 8 km 30 m − 3 km 480 m

5. (a) 4 kg 920 g + 125 g   (b) 3 kg 760 g + 4 kg 350 g
   (c) 6 kg − 4 kg 820 g   (d) 4 kg 25 g − 2 kg 230 g

6. (a) The total weight of the fruits is ▮ g.
   (b) If the apple weighs 90 g, find the
        total weight of the two pears.
   (c) If the pears are of the same weight,
        find the weight of each pear.

Capacity

 Liters and Milliliters

How much water is there in each of these beakers?

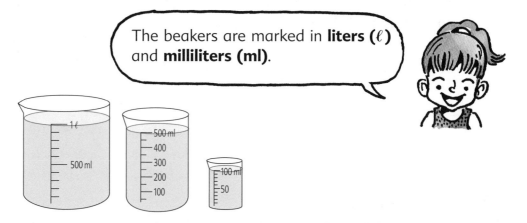

The beakers are marked in **liters (ℓ)** and **milliliters (ml)**.

What is the total amount of water in each set of beakers?

(a)

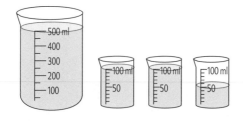

(b)

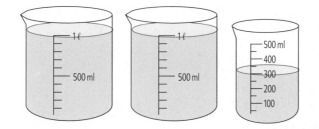

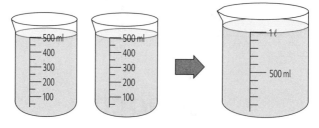

1 ℓ = 1000 ml

1. Get some paper cups.
   Find out how many paper cups you
   can fill with 1 liter of water.

2. Get a bucket.
   Find out how much water the bucket
   can hold.

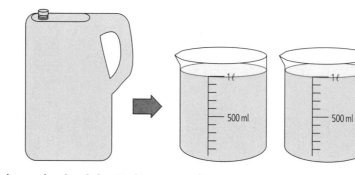

3.

The bottle holds 2 liters of water.

Its **capacity** is ▮ liters.

Workbook Exercise 17

The capacity of a container is the amount it can hold.

We measure capacity in liters and milliliters.

4. (a)

The capacity of the mug is ▮ ml.

(b)

The capacity of the bottle is ▮ ml.

(c)

The capacity of the jug is ▮ ℓ ▮ ml.

5. (a) Get a bottle which holds less than 1 liter of water.
       Estimate and then measure its capacity.
   (b) Get a bottle which holds more than 1 liter of water.
       Estimate and then measure its capacity.

Workbook Exercises 18 & 19

6. Find the total amount of water in these two beakers.

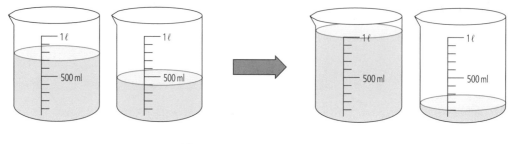

   700 ml + 400 ml = ▊ ml
                   = ▊ ℓ ▊ ml

7. Write 1500 ml in liters and milliliters.

   1500 ml = ▊ ℓ ▊ ml

8. Write in liters and milliliters.
   (a) 1200 ml   (b) 2500 ml   (c) 2050 ml
   (d) 1005 ml   (e) 3400 ml   (f) 3105 ml

9. (a) Write 2 ℓ in milliliters.
      2 ℓ = ▊ ml
   (b) Write 2 ℓ 350 ml in milliliters.
      2 ℓ 350 ml = ▊ ml

10. Write in milliliters.
    (a) 1 ℓ 800 ml   (b) 1 ℓ 80 ml   (c) 1 ℓ 8 ml
    (d) 3 ℓ 25 ml   (e) 2 ℓ 5 ml   (f) 3 ℓ 500 ml

11.

Each carton contains 250 ml of milk.

The total amount of milk in 5 cartons is ▮ ℓ ▮ ml.

12.

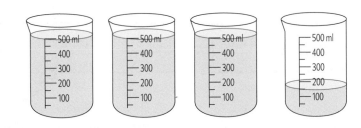

How many milliliters more water are needed to make up
2 liters?

13.

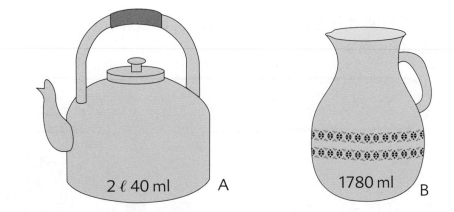

Which container can hold more water?
How much more?

Workbook Exercises 20 & 21

14.

1 ℓ 800 ml
A

3 ℓ 350 ml
B

(a) Find the total capacity of the two containers.

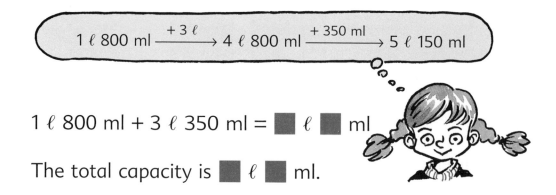

1 ℓ 800 ml $\xrightarrow{+3\,\ell}$ 4 ℓ 800 ml $\xrightarrow{+350\,ml}$ 5 ℓ 150 ml

1 ℓ 800 ml + 3 ℓ 350 ml = ⬛ ℓ ⬛ ml

The total capacity is ⬛ ℓ ⬛ ml.

(b) Find the difference in capacity between the two containers.

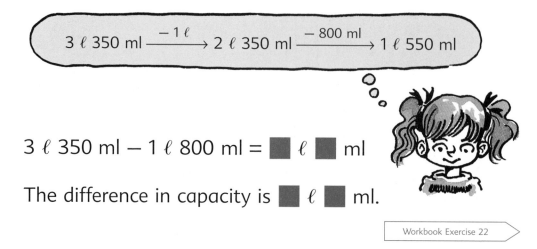

3 ℓ 350 ml $\xrightarrow{-1\,\ell}$ 2 ℓ 350 ml $\xrightarrow{-800\,ml}$ 1 ℓ 550 ml

3 ℓ 350 ml − 1 ℓ 800 ml = ⬛ ℓ ⬛ ml

The difference in capacity is ⬛ ℓ ⬛ ml.

Workbook Exercise 22

PRACTICE 4A

1. Write in milliliters.
   (a) 3 ℓ
   (b) 1 ℓ 200 ml
   (c) 2 ℓ 55 ml
   (d) 2 ℓ 650 ml
   (e) 3 ℓ 65 ml
   (f) 4 ℓ 5 ml

2. Write in liters and milliliters.
   (a) 5000 ml
   (b) 1600 ml
   (c) 2250 ml
   (d) 3205 ml
   (e) 2074 ml
   (f) 1009 ml

3. Circle the correct answer.
   (a) 1 ℓ is more than/equal to/less than 980 ml.
   (b) 2 ℓ 50 ml is more than/equal to/less than 2050 ml.
   (c) 4 ℓ 8 ml is more than/equal to/less than 4800 ml.

4. Add or subtract.
   (a) 1 ℓ 500 ml + 500 ml
   (b) 2 ℓ 800 ml + 1 ℓ 200 ml
   (c) 3 ℓ 300 ml + 750 ml
   (d) 5 ℓ 900 ml + 3 ℓ 240 ml
   (e) 2 ℓ 800 ml − 1 ℓ 780 ml
   (f) 4 ℓ − 1 ℓ 850 ml
   (g) 4 ℓ 80 ml − 1 ℓ 360 ml
   (h) 6 ℓ 5 ml − 2 ℓ 80 ml

5. The table shows the capacities of four containers.

Container A	2 ℓ 375 ml
Container B	1 ℓ 750 ml
Container C	1755 ml
Container D	2150 ml

   (a) Which container has the greatest capacity?
   (b) Which container has the smallest capacity?
   (c) What is the total capacity of the four containers?

PRACTICE 4B

1. The capacity of Container A is 2 ℓ 650 ml.
   The capacity of Container B is 5 ℓ 300 ml.
   (a) What is the total capacity of the two containers?
   (b) How much more water can Container B hold than
       Container A?

2. Container X holds 2 ℓ 800 ml of water.
   Container Y holds 1 ℓ 600 ml more water than Container X.
   How much water does Container Y hold?

3. Mrs. Chen fills a container with 9 cartons of orange juice.
   Each carton contains 2 liters of orange juice.
   What is the capacity of the container?

4. The capacity of a bucket is 6 liters.
   5 buckets of water are needed to fill up a tank.
   What is the capacity of the tank?

5. The capacity of a container is 24 liters.
   How many buckets of water are needed to fill up the
   container if the capacity of the bucket is 3 liters?

6. The capacity of a container is 8 liters.
   It contains 4 ℓ 650 ml of water.
   How much more water is needed to fill up the container?

7. Adam bought 6 cans of paint.
   Each can contained 3 liters of paint.
   He had 2 ℓ 400 ml of paint left after painting his house.
   How much paint did he use?

2 **Gallons, Quarts, Pints and Cups**

How much water is there in each of these containers?

Remember?
1 gallon (gal) = 4 quarts (qt)
1 quart = 2 pints (pt)
1 pint = 2 cups (c)

1 gal   1 qt   1 pt   1 c

1. Look at some measuring cups.
    Can you find the marking for quarts, pints and cups?

2.

This jug holds 1 quart of water.
Its capacity is 1 quart.

1 quart of water = 2 pints of water.
We can also say the capacity of this jug is 2 pints.

Do you know the capacity of this jug in cups?

3. Find the total amount of water in these two containers.

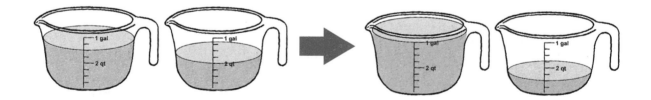

3 qt + 2 qt = ▮ qt

= ▮ gal ▮ qt

4. Write 78 qt in gallons and quarts.

78 qt = ▮ gal ▮ qt

5. Write 15 pt in quarts and pints.

1 qt = 2 pt

6. Write 21 c in pints and cups.

7.

Each carton contains 1 pt of milk.
The total amount of milk in the five cartons is ▨ qt ▨ pt.

8.

2 pt 1 c

3 pt 1 c

(a) Find the total amount of milk in the two jugs.

2 pt 1 c + 3 pt 1 c = ▨ pt ▨ c

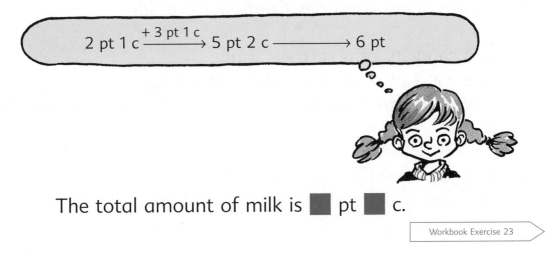

2 pt 1 c $\xrightarrow{+\ 3\ pt\ 1\ c}$ 5 pt 2 c $\longrightarrow$ 6 pt

The total amount of milk is ▨ pt ▨ c.

Workbook Exercise 23

55

PRACTICE 4C

1. Write in cups.
   (a) 8 pt
   (b) 15 pt 1 c

2. Write in pints.
   (a) 7 qt
   (b) 11 qt 1 pt

3. Write in quarts.
   (a) 10 gal
   (b) 23 gal 1 qt

4. Fill in the blanks with **>** (greater than), **=** (equal to), or **<** (less than).
   (a) 5 gal 1 qt ■ 12 qt
   (b) 12 qt 1 pt ■ 23 pt
   (c) 15 c ■ 7 pt 1 c

5. Add or subtract.
   (a) 3 qt 1 pt + 7 qt 1 pt = ■
   (b) 12 gal − 7 gal 1 qt = ■
   (c) 258 pt − 185 pt 1 c = ■

6. Container A holds 13 gal water.
   Container B holds 7 gal 1 qt less water than Container A.
   How much water does Container B hold?

7. The capacity of a kettle is 2 qt.
   There is 1 pt of water in the kettle now.
   How many more pints of water are needed to fill up the kettle?

8. Morgan drinks 2 c of milk daily.
   How many pints of milk does she drink in a week?

REVIEW C

Find the value of each of the following:

	(a)	(b)	(c)
1.	895 + 5037	6409 + 399	2846 + 754
2.	1436 − 437	3002 − 78	5362 − 26
3.	77 × 4	73 × 9	123 × 5
4.	900 ÷ 2	408 ÷ 3	518 ÷ 8

5. Cameron received $504 for selling pens at $8 each.
   How many pens did he sell?

6. Eric paid $628 for a television set and $1485 for
   a computer.
   He had $515 left.
   How much money did he have at first?

7. Mrs. Ray bought 4 boxes of cookies.
   There were 12 chocolate cookies and 8 sugar cookies in
   each box.
   How many cookies were there altogether?

8. Kyle sold 337 boxes of cookies last month.
   He sold 299 more boxes this month than last month.
   How many boxes of cookies did he sell in the two months?

9. Sally bought 200 eggs to make cakes.
   She used 8 eggs to make each cake.
   (a) How many cakes did she make?
   (b) If she sold all the cakes at $10 each, how much money
       would she receive?

5

Graphs

1 Bar Graphs

This picture graph shows the number of fish caught by four boys.

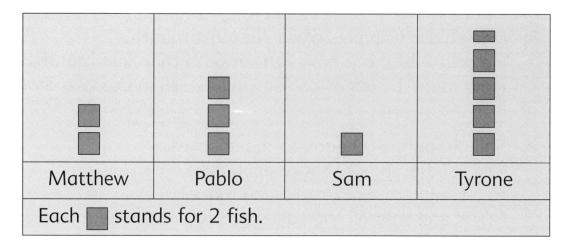

Matthew	Pablo	Sam	Tyrone

Each ■ stands for 2 fish.

How many fish did each boy catch?

This **bar graph** shows the same information.

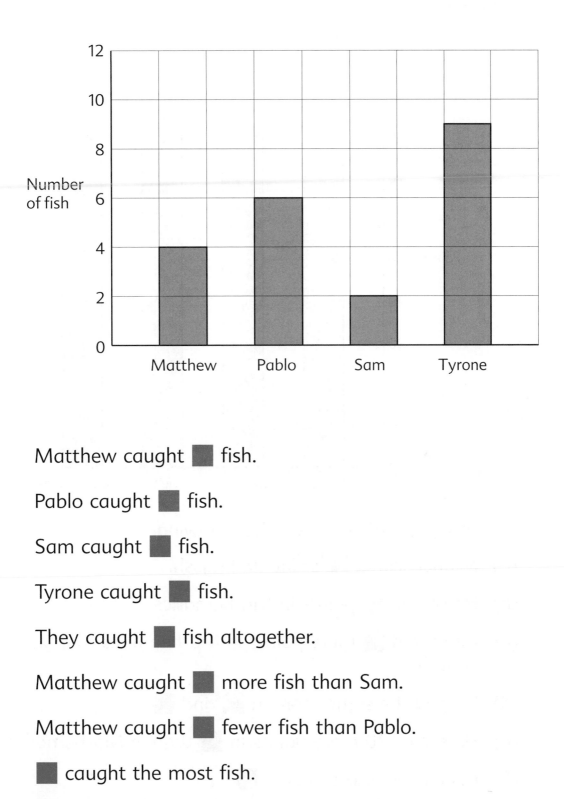

Matthew caught ■ fish.

Pablo caught ■ fish.

Sam caught ■ fish.

Tyrone caught ■ fish.

They caught ■ fish altogether.

Matthew caught ■ more fish than Sam.

Matthew caught ■ fewer fish than Pablo.

■ caught the most fish.

■ caught the fewest fish.

59

1. This bar graph shows Miguel's examination results for four subjects.

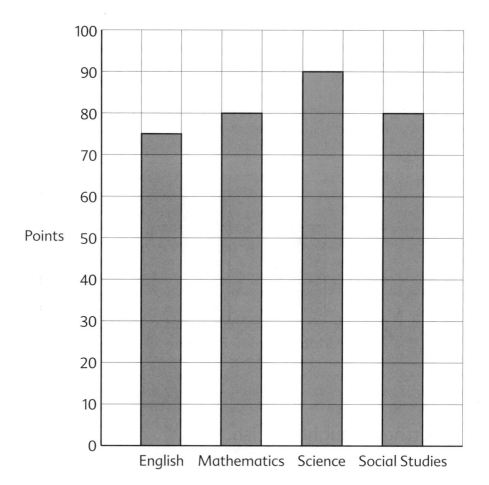

Use the graph to complete the following.

(a) Miguel scored 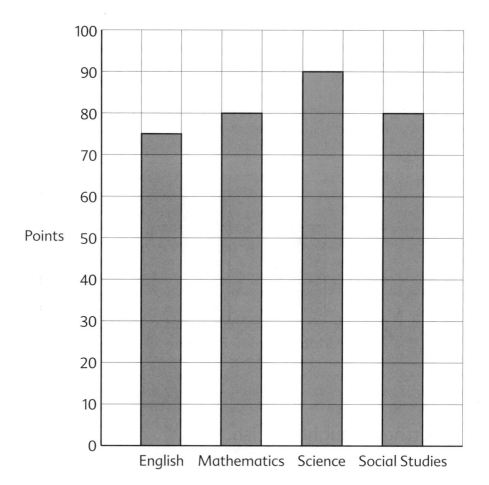 points in English.

(b) He scored ▪ points in Mathematics.

(c) He scored ▪ more points in Mathematics than in English.

(d) He had the same score in ▪ and ▪.

(e) He scored 10 more points in ▪ than in Mathematics.

(f) His highest score was in ▪.

(g) His lowest score was in ▪.

2. This bar graph shows Sally's savings for four months.

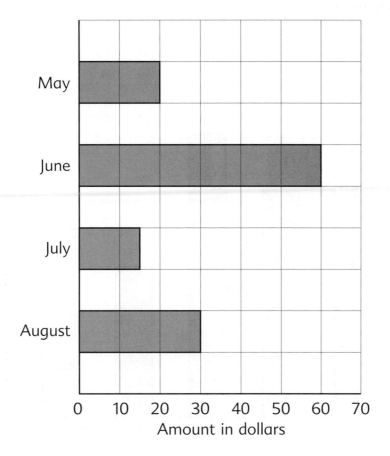

Amount in dollars

Use the graph to answer the following questions.

(a) How much did Sally save in May?

(b) How much more did she save in June than in May?

(c) In which month did she save $15?

(d) In which month did she save the most?

(e) In which month did she save twice as much as in August?

(f) What was her total savings for the four months?

Workbook Exercise 24

3. This bar graph shows the number of books read by five children in a year.

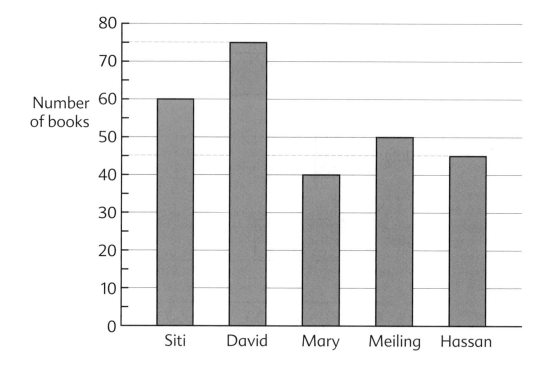

Use the graph to answer the following questions.

(a) How many books did Siti read?

(b) How many books did David read?

(c) How many more books did David read than Meiling?

(d) Who read 5 fewer books than Meiling?

(e) Who read the most books?

(f) Who read the fewest books?

4. This bar graph shows the number of people who visited a book fair from Monday to Friday.

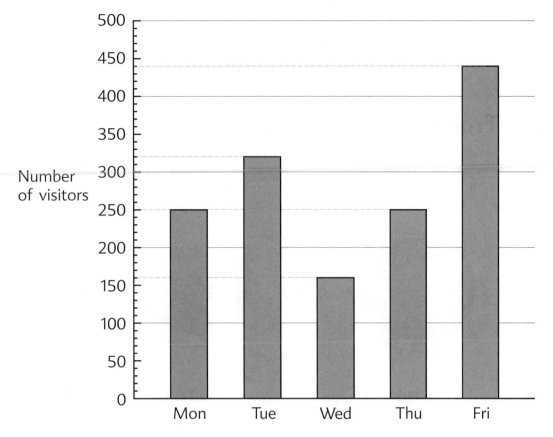

Use the graph to answer the following questions.

(a) How many people visited the book fair on Tuesday?

(b) How many more people visited the book fair on Friday than on Thursday?

(c) On which day was the number of visitors the smallest?

(d) On which day were there as many visitors as on Monday?

(e) On which day were there twice as many visitors as on Wednesday?

(f) If there were 200 adults on Tuesday, how many children were there?

Workbook Exercise 25

Fractions

1 Fraction of a Whole

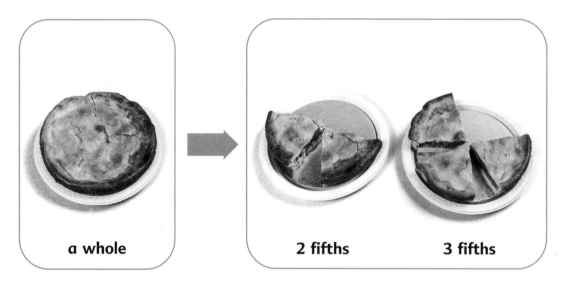

| a whole | 2 fifths | 3 fifths |

How many fifths are there in a whole?

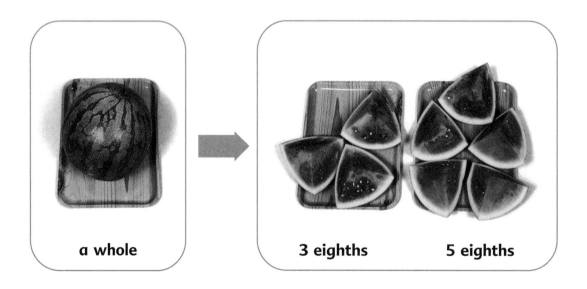

| a whole | 3 eighths | 5 eighths |

How many eighths are there in a whole?

1.

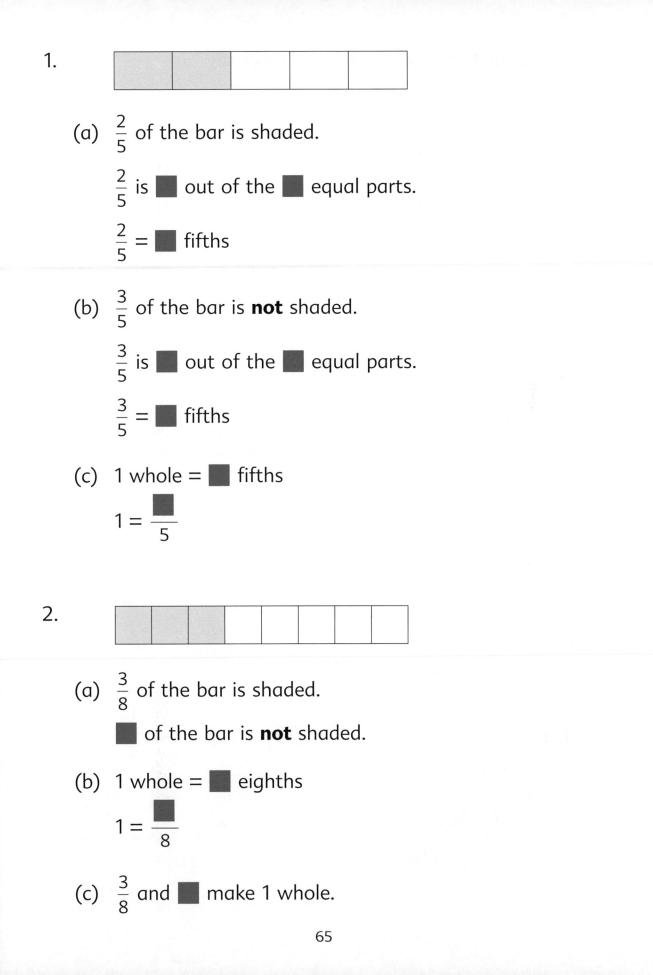

(a) $\frac{2}{5}$ of the bar is shaded.

$\frac{2}{5}$ is ■ out of the ■ equal parts.

$\frac{2}{5}$ = ■ fifths

(b) $\frac{3}{5}$ of the bar is **not** shaded.

$\frac{3}{5}$ is ■ out of the ■ equal parts.

$\frac{3}{5}$ = ■ fifths

(c) 1 whole = ■ fifths

$1 = \dfrac{■}{5}$

2.

(a) $\frac{3}{8}$ of the bar is shaded.

■ of the bar is **not** shaded.

(b) 1 whole = ■ eighths

$1 = \dfrac{■}{8}$

(c) $\frac{3}{8}$ and ■ make 1 whole.

3. What fraction of each shape is shaded?

(a)

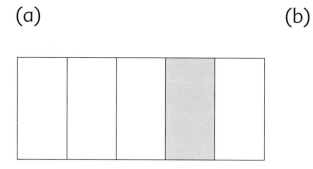

(b)

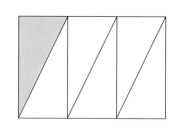

(c)

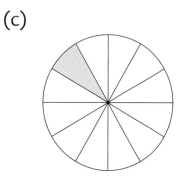

(d)

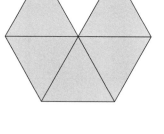

(e)

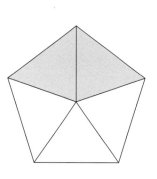

(f)

(g)

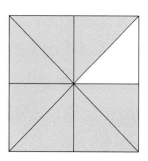

(h)

Workbook Exercises 26 to 28

4.

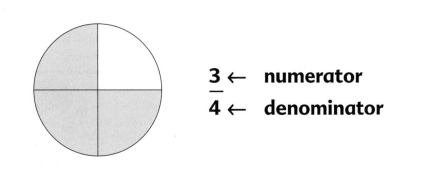

In the fraction $\frac{3}{4}$, 3 is the numerator and 4 is the denominator.

Name the numerator and denominator of each of these fractions.

(a) $\frac{2}{5}$ (b) $\frac{4}{10}$ (c) $\frac{6}{7}$ (d) $\frac{6}{9}$

5. Which is greater, $\frac{1}{5}$ or $\frac{1}{3}$?

6. Which is greater, $\frac{3}{4}$ or $\frac{3}{5}$?

7. Which is greater, $\frac{3}{8}$ or $\frac{5}{8}$?

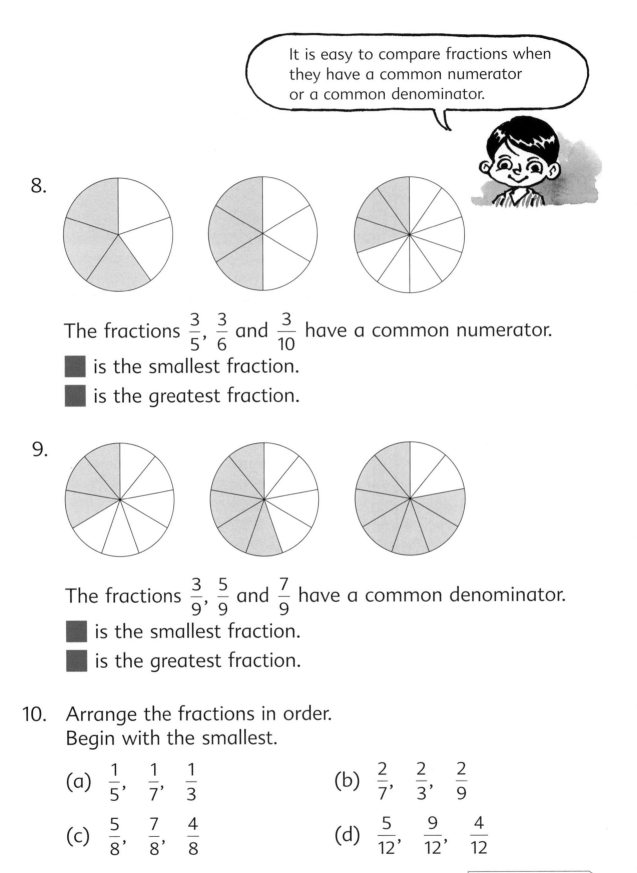

It is easy to compare fractions when they have a common numerator or a common denominator.

8.

The fractions $\frac{3}{5}$, $\frac{3}{6}$ and $\frac{3}{10}$ have a common numerator.

▇ is the smallest fraction.

▇ is the greatest fraction.

9.

The fractions $\frac{3}{9}$, $\frac{5}{9}$ and $\frac{7}{9}$ have a common denominator.

▇ is the smallest fraction.

▇ is the greatest fraction.

10. Arrange the fractions in order.
Begin with the smallest.

(a) $\frac{1}{5}$, $\frac{1}{7}$, $\frac{1}{3}$

(b) $\frac{2}{7}$, $\frac{2}{3}$, $\frac{2}{9}$

(c) $\frac{5}{8}$, $\frac{7}{8}$, $\frac{4}{8}$

(d) $\frac{5}{12}$, $\frac{9}{12}$, $\frac{4}{12}$

Workbook Exercises 29 & 30

PRACTICE 6A

1. Find the missing numbers.

    (a) $\dfrac{1}{4}$ and 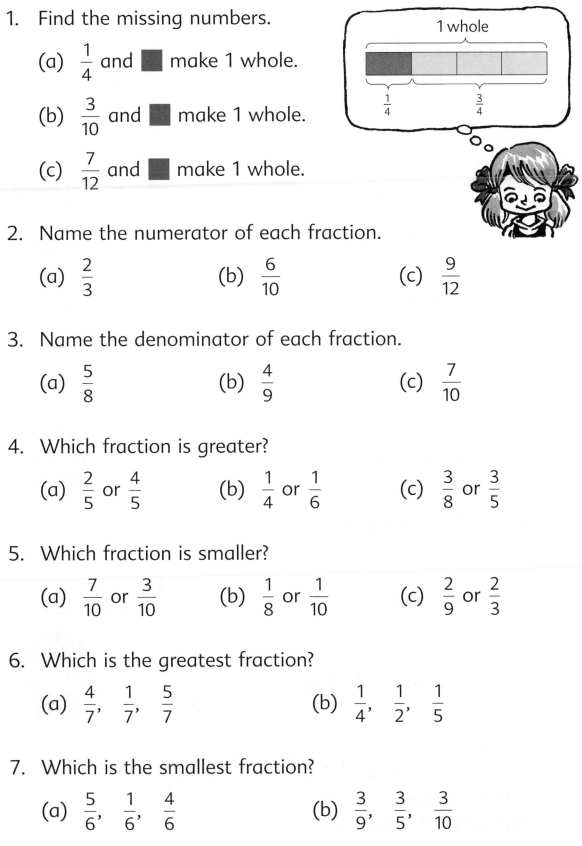 make 1 whole.

    (b) $\dfrac{3}{10}$ and ▮ make 1 whole.

    (c) $\dfrac{7}{12}$ and ▮ make 1 whole.

2. Name the numerator of each fraction.

    (a) $\dfrac{2}{3}$ (b) $\dfrac{6}{10}$ (c) $\dfrac{9}{12}$

3. Name the denominator of each fraction.

    (a) $\dfrac{5}{8}$ (b) $\dfrac{4}{9}$ (c) $\dfrac{7}{10}$

4. Which fraction is greater?

    (a) $\dfrac{2}{5}$ or $\dfrac{4}{5}$ (b) $\dfrac{1}{4}$ or $\dfrac{1}{6}$ (c) $\dfrac{3}{8}$ or $\dfrac{3}{5}$

5. Which fraction is smaller?

    (a) $\dfrac{7}{10}$ or $\dfrac{3}{10}$ (b) $\dfrac{1}{8}$ or $\dfrac{1}{10}$ (c) $\dfrac{2}{9}$ or $\dfrac{2}{3}$

6. Which is the greatest fraction?

    (a) $\dfrac{4}{7}, \dfrac{1}{7}, \dfrac{5}{7}$ (b) $\dfrac{1}{4}, \dfrac{1}{2}, \dfrac{1}{5}$

7. Which is the smallest fraction?

    (a) $\dfrac{5}{6}, \dfrac{1}{6}, \dfrac{4}{6}$ (b) $\dfrac{3}{9}, \dfrac{3}{5}, \dfrac{3}{10}$

2 Equivalent Fractions

Fold a piece of paper into 2 equal parts.
Shade 1 part.

1 out of
2 equal parts

$\frac{1}{2}$ of the paper is shaded.

Fold the paper again.

2 out of
4 equal parts

$\frac{2}{4}$ of the paper is shaded.

Fold the paper again.

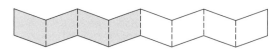

4 out of
8 equal parts

$\frac{4}{8}$ of the paper is shaded.

The fractions $\frac{1}{2}$, $\frac{2}{4}$ and $\frac{4}{8}$ have different numerators and
denominators.
But they are equal.

$$\frac{1}{2} = \frac{2}{4} = \frac{4}{8}$$

$\frac{1}{2}$, $\frac{2}{4}$ and $\frac{4}{8}$ are **equivalent fractions**.

Name some more equivalent fractions of $\frac{1}{2}$.

$\frac{2}{4}$ and $\frac{4}{8}$ are different ways of writing $\frac{1}{2}$.

1.

$\frac{2}{3}$ of the bar is shaded.

(a) $\frac{2}{3} = \frac{\blacksquare}{6}$

(b) $\frac{2}{3} = \frac{\blacksquare}{9}$

(c) $\frac{2}{3} = \frac{\blacksquare}{12}$

(d) Name some more equivalent fractions of $\frac{2}{3}$.

2. What are the missing numerators and denominators?

(a)

$$1 \quad = \quad \frac{\blacksquare}{2} \quad = \quad \frac{3}{\blacksquare} \quad = \quad \frac{\blacksquare}{\blacksquare}$$

(b)

$$\frac{1}{3} \quad = \quad \frac{\blacksquare}{6} \quad = \quad \frac{3}{\blacksquare} \quad = \quad \frac{\blacksquare}{\blacksquare}$$

> To find an equivalent fraction, multiply the numerator and denominator by the same number.

$$\frac{1}{3} \overset{\times 2}{\underset{\times 2}{=}} \frac{\blacksquare}{6} \qquad \frac{1}{3} \overset{\times 3}{\underset{\times 3}{=}} \frac{3}{\blacksquare}$$

3. Find the missing numerator or denominator.

(a) $\frac{1}{4} = \frac{\blacksquare}{12}$ (b) $\frac{2}{3} = \frac{\blacksquare}{9}$ (c) $\frac{1}{5} = \frac{\blacksquare}{10}$

(d) $\frac{1}{6} = \frac{3}{\blacksquare}$ (e) $\frac{3}{5} = \frac{6}{\blacksquare}$ (f) $\frac{3}{4} = \frac{6}{\blacksquare}$

Workbook Exercises 31 & 32

4. What are the missing numerators and denominators?

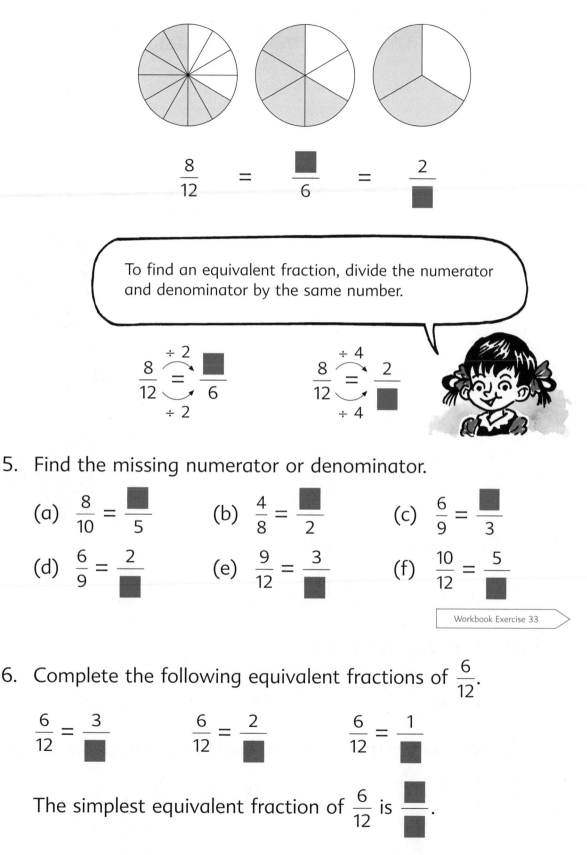

$$\frac{8}{12} = \frac{\blacksquare}{6} = \frac{2}{\blacksquare}$$

To find an equivalent fraction, divide the numerator and denominator by the same number.

$$\frac{8}{12} \overset{\div 2}{\underset{\div 2}{=}} \frac{\blacksquare}{6}$$

$$\frac{8}{12} \overset{\div 4}{\underset{\div 4}{=}} \frac{2}{\blacksquare}$$

5. Find the missing numerator or denominator.

(a) $\frac{8}{10} = \frac{\blacksquare}{5}$

(b) $\frac{4}{8} = \frac{\blacksquare}{2}$

(c) $\frac{6}{9} = \frac{\blacksquare}{3}$

(d) $\frac{6}{9} = \frac{2}{\blacksquare}$

(e) $\frac{9}{12} = \frac{3}{\blacksquare}$

(f) $\frac{10}{12} = \frac{5}{\blacksquare}$

Workbook Exercise 33

6. Complete the following equivalent fractions of $\frac{6}{12}$.

$$\frac{6}{12} = \frac{3}{\blacksquare} \qquad \frac{6}{12} = \frac{2}{\blacksquare} \qquad \frac{6}{12} = \frac{1}{\blacksquare}$$

The simplest equivalent fraction of $\frac{6}{12}$ is $\frac{\blacksquare}{\blacksquare}$.

7. Express each of the following fractions in its simplest form.

(a) $\dfrac{2}{4}$

(b) $\dfrac{6}{8}$

(c) $\dfrac{5}{10}$

(d) $\dfrac{3}{9}$

(e) $\dfrac{4}{12}$

(f) $\dfrac{4}{6}$

(g) $\dfrac{10}{12}$

(h) $\dfrac{6}{10}$

Workbook Exercise 34

8. Which is greater, $\dfrac{3}{4}$ or $\dfrac{5}{8}$?

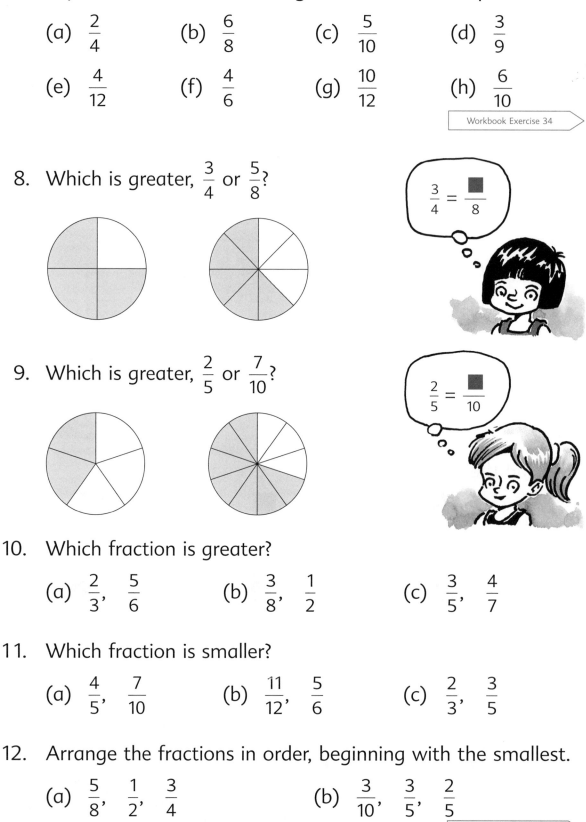

$\dfrac{3}{4} = \dfrac{\blacksquare}{8}$

9. Which is greater, $\dfrac{2}{5}$ or $\dfrac{7}{10}$?

$\dfrac{2}{5} = \dfrac{\blacksquare}{10}$

10. Which fraction is greater?

(a) $\dfrac{2}{3}$, $\dfrac{5}{6}$

(b) $\dfrac{3}{8}$, $\dfrac{1}{2}$

(c) $\dfrac{3}{5}$, $\dfrac{4}{7}$

11. Which fraction is smaller?

(a) $\dfrac{4}{5}$, $\dfrac{7}{10}$

(b) $\dfrac{11}{12}$, $\dfrac{5}{6}$

(c) $\dfrac{2}{3}$, $\dfrac{3}{5}$

12. Arrange the fractions in order, beginning with the smallest.

(a) $\dfrac{5}{8}$, $\dfrac{1}{2}$, $\dfrac{3}{4}$

(b) $\dfrac{3}{10}$, $\dfrac{3}{5}$, $\dfrac{2}{5}$

Workbook Exercise 35

PRACTICE 6B

1. Find the missing numerator in each of the following:

   (a) $\dfrac{1}{4} = \dfrac{\blacksquare}{8}$

   (b) $\dfrac{3}{5} = \dfrac{\blacksquare}{15}$

   (c) $\dfrac{1}{3} = \dfrac{\blacksquare}{6} = \dfrac{\blacksquare}{9}$

   (d) $\dfrac{4}{10} = \dfrac{\blacksquare}{5}$

   (e) $\dfrac{6}{9} = \dfrac{\blacksquare}{3}$

   (f) $\dfrac{1}{2} = \dfrac{\blacksquare}{4} = \dfrac{\blacksquare}{6}$

2. Find the missing denominator in each of the following:

   (a) $\dfrac{2}{5} = \dfrac{4}{\blacksquare}$

   (b) $\dfrac{3}{4} = \dfrac{9}{\blacksquare}$

   (c) $\dfrac{2}{3} = \dfrac{4}{\blacksquare} = \dfrac{6}{\blacksquare}$

   (d) $\dfrac{6}{12} = \dfrac{1}{\blacksquare}$

   (e) $\dfrac{6}{8} = \dfrac{3}{\blacksquare}$

   (f) $\dfrac{1}{2} = \dfrac{3}{\blacksquare} = \dfrac{5}{\blacksquare}$

3. Circle the greater fraction.

   (a) $\dfrac{3}{10}, \quad \dfrac{7}{10}$

   (b) $\dfrac{5}{6}, \quad \dfrac{9}{12}$

   (c) $\dfrac{10}{12}, \quad \dfrac{4}{5}$

   (d) $\dfrac{1}{2}, \quad \dfrac{5}{6}$

   (e) $\dfrac{3}{4}, \quad \dfrac{2}{3}$

   (f) $\dfrac{3}{5}, \quad \dfrac{5}{8}$

4. Arrange the fractions in order, beginning with the smallest fraction.

   (a) $\dfrac{3}{7}, \quad \dfrac{1}{7}, \quad \dfrac{5}{7}$

   (b) $\dfrac{1}{5}, \quad \dfrac{1}{2}, \quad \dfrac{1}{10}$

   (c) $\dfrac{2}{3}, \quad \dfrac{1}{2}, \quad \dfrac{5}{6}$

   (d) $\dfrac{2}{3}, \quad \dfrac{1}{4}, \quad \dfrac{5}{12}$

5. Melissa ate $\dfrac{2}{6}$ of a pie.

   Sara ate $\dfrac{1}{2}$ of the pie.

   Who ate a bigger portion of the pie?

REVIEW D

1. Write the numbers.
   (a) Nine thousand, two hundred ten
   (b) Four thousand, sixty

2. Write the numbers in words.
   (a) 6204   (b) 3540   (c) 5028

3. What number is 100 less than 4000?

4. Write the numbers in order, beginning with the smallest.
   (a) 4104, 4014, 4041, 4410
   (b) 2211, 1112, 2111, 2121

5. Find the product of 125 and 8.

6. Find the quotient and remainder when 500 is divided by 8.

7. What number must be subtracted from 55 to give the answer 44?

8. (a) How many $10 bills can you get for $200?
   (b) How many 5¢ coins can you get for $1.50?

9. Find the missing numerator or denominator.

   (a) $\dfrac{1}{4} = \dfrac{\blacksquare}{12}$ (b) $\dfrac{2}{3} = \dfrac{6}{\blacksquare}$ (c) $\dfrac{8}{10} = \dfrac{4}{\blacksquare}$

10. Circle the smaller fraction.

    (a) $\dfrac{1}{3}, \dfrac{1}{4}$ (b) $\dfrac{2}{7}, \dfrac{4}{7}$ (c) $\dfrac{10}{10}, \dfrac{11}{12}$

    (d) $\dfrac{3}{6}, \dfrac{5}{8}$ (e) $\dfrac{2}{5}, \dfrac{3}{8}$ (f) $\dfrac{3}{7}, \dfrac{2}{5}$

11. Find the missing numbers.

(a) 4 m 20 cm = ■ cm

(b) 205 cm = ■ m ■ cm

(c) 2 km 95 m = ■ m

(d) 1600 m = ■ km ■ m

(e) 1 kg 40 g = ■ g

(f) 2450 g = ■ kg ■ g

(g) 3 ℓ 60 ml = ■ ml

(h) 2525 ml = ■ ℓ ■ ml

12. John bought 6 pears.
He gave the cashier $5.
How much change did he receive?

PEARS
70¢ each

13. A badminton racket costs $9.60.
A tennis racket costs $38.40.
How much more is the cost of the tennis racket than the cost of the badminton racket?

14. The tub contains 1 liter of ice cream.
Peter and his friends eat 325 ml of it.
How much ice cream is left?

15. Sally bought 10 cartons of milk.
Each carton contained 125 ml of milk.
Find the total amount of milk in liters and milliliters.

16. Juan spent $\frac{4}{9}$ of his allowance and saved the rest.
What fraction of his allowance did he save?

17. Matthew spent $\frac{3}{7}$ of his money on a book and the rest on a racket.
What fraction of his money was spent on the racket?

7
Time

1 Hours and Minutes

How far is 3 km?
How long would I take to run 3 km?

David

8:20
20 minutes past 8

8:35
25 minutes to 9

David started running at 8:20 a.m.
He ran 3 km.
He finished at 8:35 a.m.
He took 15 minutes to run 3 km.

We read 8:20 as **eight twenty**. 8:20 is 20 minutes after 8 o'clock. We say the time is twenty minutes past eight.

We read 8:35 as **eight thirty-five**. 8:35 is 25 minutes before 9 o'clock. We say the time is twenty-five minutes to nine.

1. Find out how many times you can write your name in 1 minute.

2. What time is it?

(a)

(b)

(c)

(d)

(e)

(f)

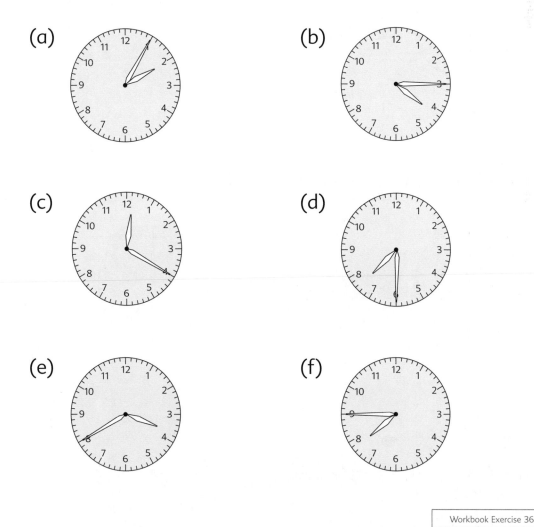

Workbook Exercise 36

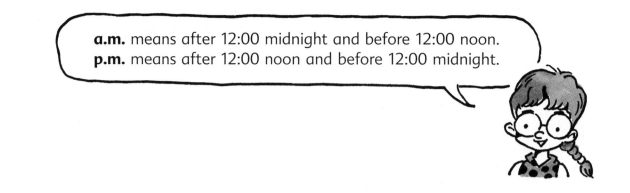

a.m. means after 12:00 midnight and before 12:00 noon.
p.m. means after 12:00 noon and before 12:00 midnight.

3. What time is 26 minutes after 9:30 a.m.?

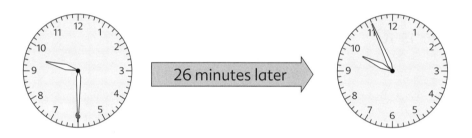

26 minutes later

4. How many minutes are there in **1 hour**?

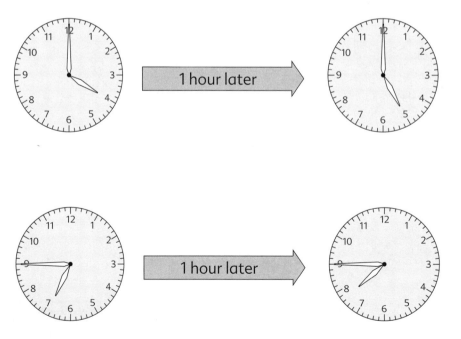

1 hour later

1 hour later

5. (a) How many **minutes** are there from 1:15 p.m. to 1:42 p.m.?

(b) How many **hours** are there from 3:18 p.m. to 8:18 p.m.?

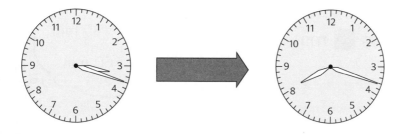

(c) How long is it from 9:15 a.m. to 11:30 a.m.?

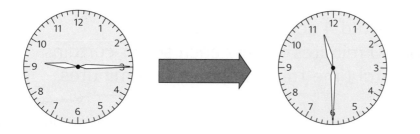

The **hour (h)** and **minute (min)** are units of time.

1 hour = 60 minutes

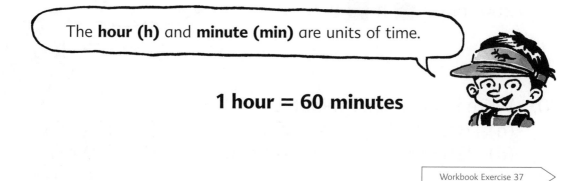

Workbook Exercise 37

6. The table shows the time taken by 3 children to paint a picture.

   (a) Who took the longest time?
   (b) Who took the shortest time?

Name	Time taken
Amy	1 h 15 min
Jane	2 h 5 min
Siti	1 h 20 min

7. Mary took 1 h 35 min to complete a jigsaw puzzle. Write the time taken in minutes.

   1 h 35 min = 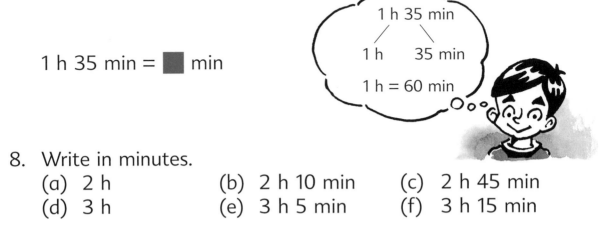 min

8. Write in minutes.
   (a) 2 h   (b) 2 h 10 min   (c) 2 h 45 min
   (d) 3 h   (e) 3 h 5 min   (f) 3 h 15 min

9. Mrs. Lee sewed 4 sets of curtains.
   She took 50 minutes to sew each set of curtains.
   Find the total time taken in hours and minutes.

   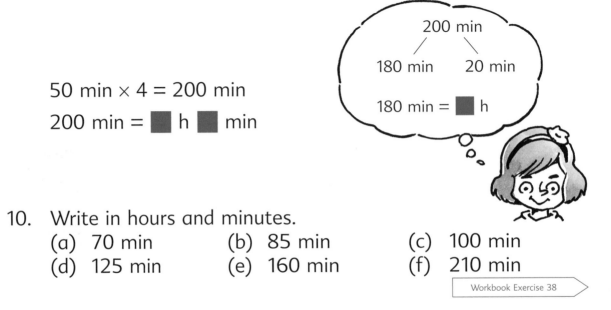

   50 min × 4 = 200 min

   200 min = ▮ h ▮ min

10. Write in hours and minutes.
    (a) 70 min   (b) 85 min   (c) 100 min
    (d) 125 min   (e) 160 min   (f) 210 min

    Workbook Exercise 38

82

11. A plane left San Francisco at 8:00 a.m.
    It arrived in Portland at 9:05 a.m.
    How long did the journey take?

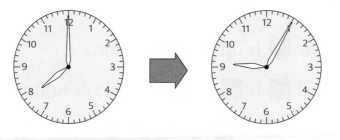

The journey took ■ h ■ min.

12. Kristi went to market at 7:15 a.m.
    She came home 1 h 45 min later.
    When did she come home?

What time is
1 h 45 min
after 7:15 a.m.?

She came home at ■ a.m.

13. Sally took 1 h 10 min to do her homework.
    She finished doing her homework at 9:40 p.m.
    When did she start?

What time is
1 h 10 min
before 9:40 p.m.?

She started at ■ p.m.

Workbook Exercise 39

14.

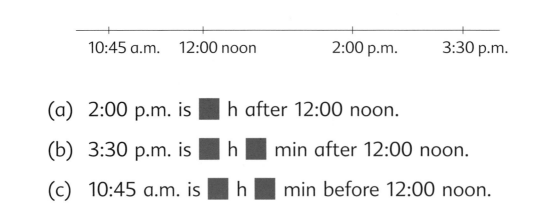

(a) 2:00 p.m. is ■ h after 12:00 noon.

(b) 3:30 p.m. is ■ h ■ min after 12:00 noon.

(c) 10:45 a.m. is ■ h ■ min before 12:00 noon.

15. A supermarket is open from 10:15 a.m. to 9:30 p.m. every day.
How long is the supermarket open a day?

```
                                ?
    ┌─────────────────────────────────────────────────────┐
10:15 a.m.  12:00 noon                            9:30 p.m.
    └──┬──────┴──────────────────────────────────────┘
  1 h 45 min              9 h 30 min
```

1 h 45 min $\xrightarrow{+\,9\,h}$ 10 h 45 min $\xrightarrow{+\,30\,min}$ 11 h 15 min

1 h 45 min + 9 h 30 min = ■ h ■ min

The supermarket is open ■ h ■ min a day.

84

16.

| 9:10 p.m. | 12:00 midnight | 4:00 a.m. | 6:40 a.m. |

(a) 4:00 a.m. is ■ h after 12:00 midnight.

(b) 6:40 a.m. is ■ h ■ min after 12:00 midnight.

(c) 9:10 p.m. is ■ h ■ min before 12:00 midnight.

17. A night tour began at 10:30 p.m. and lasted 3 h 20 min. When did the night tour end?

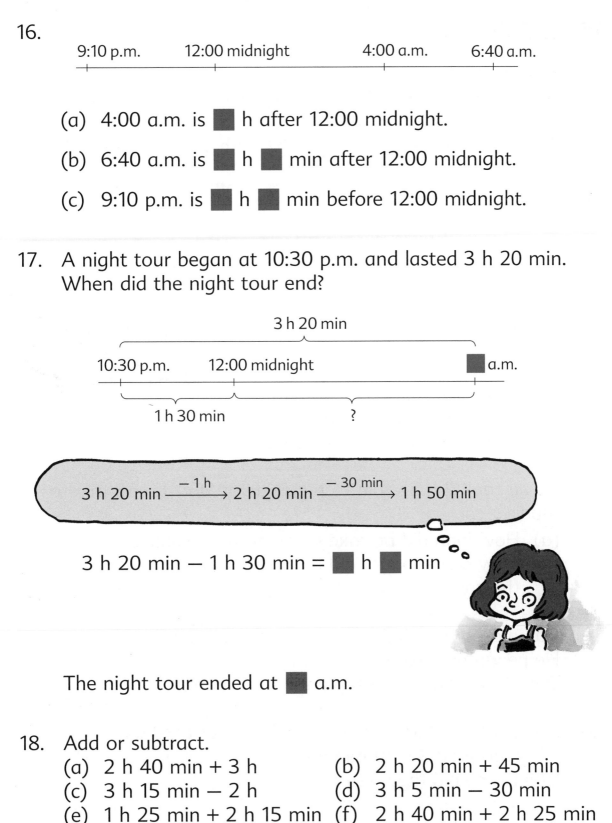

3 h 20 min

10:30 p.m.   12:00 midnight   ■ a.m.

1 h 30 min   ?

3 h 20 min —− 1 h→ 2 h 20 min —− 30 min→ 1 h 50 min

3 h 20 min − 1 h 30 min = ■ h ■ min

The night tour ended at ■ a.m.

18. Add or subtract.
   (a) 2 h 40 min + 3 h   (b) 2 h 20 min + 45 min
   (c) 3 h 15 min − 2 h   (d) 3 h 5 min − 30 min
   (e) 1 h 25 min + 2 h 15 min (f) 2 h 40 min + 2 h 25 min
   (g) 3 h 50 min − 1 h 35 min (h) 3 h 20 min − 1 h 40 min

Workbook Exercise 40

PRACTICE 7A

1. Add or subtract.
   (a) 1 h 45 min + 2 h   (b) 3 h 40 min − 2 h
   (c) 2 h 15 min + 45 min   (d) 3 h − 1 h 45 min
   (e) 1 h 30 min + 1 h 50 min (f) 2 h 10 min − 1 h 30 min

2. This clock is 5 minutes slow.
   What is the correct time?

3. How long is it?
   (a) From 4:40 a.m. to 11:55 a.m.
   (b) From 5:45 p.m. to 7:00 p.m.
   (c) From 10:05 p.m. to 12:00 midnight
   (d) From 2:40 p.m. to 3:25 p.m.

4. David took 2 h 35 min to repair a van and 1 h 55 min to repair a car.
   (a) How long did he take to repair both vehicles?
   (b) How much longer did he take to repair the van than the car?

5. Cameron took 2 h 30 min to paint his room.
   He began at 9:20 a.m.
   What time did he finish painting his room?

6. A group of children left for a field trip at 8:30 a.m.
   They returned 4 h 10 min later.
   What time did they return?

7. A supermarket opens for business at 9:30 a.m.
   Its workers have to report for work 40 minutes earlier.
   What time must the workers report for work?

2 Other Units of Time

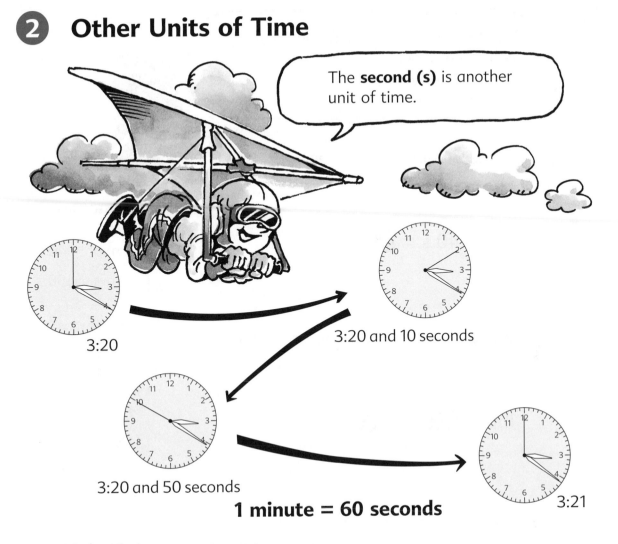

The **second (s)** is another unit of time.

3:20

3:20 and 10 seconds

3:20 and 50 seconds

1 minute = 60 seconds

3:21

1. (a) Find out how many times you can skip in 10 seconds.

   (b) How long do you take to write the word CHILDREN?
   (c) How long do you take to run 100 m?

Workbook Exercise 41

The hour (h), minute (min) and second (s) are units of time.

1 h = 60 min

1 min = 60 s

2. (a) Write 3 min 40 s in seconds.

3 min 40 s = 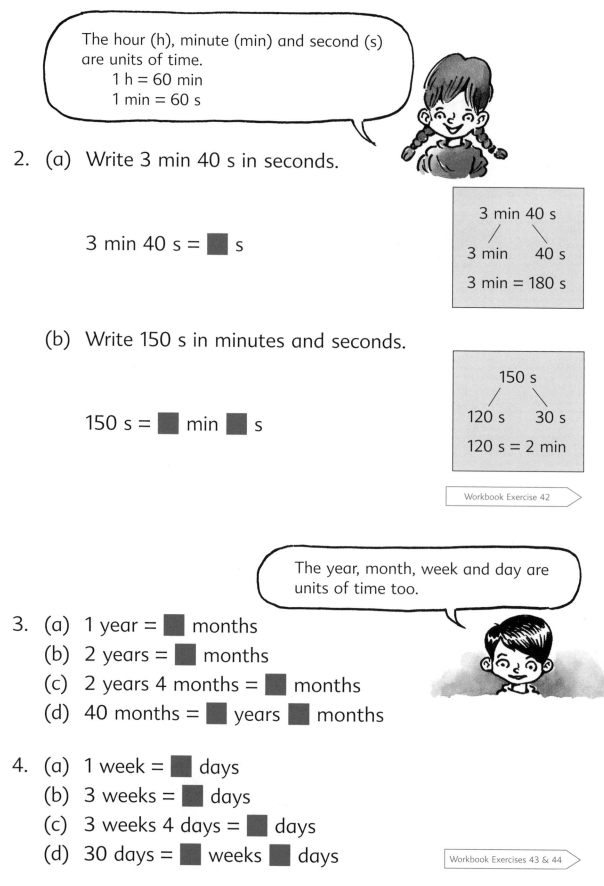 s

3 min 40 s

3 min   40 s

3 min = 180 s

(b) Write 150 s in minutes and seconds.

150 s = ■ min ■ s

150 s

120 s   30 s

120 s = 2 min

Workbook Exercise 42

The year, month, week and day are units of time too.

3. (a) 1 year = ■ months
   (b) 2 years = ■ months
   (c) 2 years 4 months = ■ months
   (d) 40 months = ■ years ■ months

4. (a) 1 week = ■ days
   (b) 3 weeks = ■ days
   (c) 3 weeks 4 days = ■ days
   (d) 30 days = ■ weeks ■ days

Workbook Exercises 43 & 44

PRACTICE 7B

1. Find the missing numbers.

    (a) 2 h 12 min = ■ min   (b) 108 min = ■ h ■ min

    (c) 2 min 3 s = ■ s   (d) 94 s = ■ min ■ s

    (e) 1 year 9 months = ■ months

    (f) 30 months = ■ years ■ months

    (g) 2 weeks 5 days = ■ days

    (h) 40 days = ■ weeks ■ days

2. The flying time from Chicago to Minneapolis is 1 h 35 min
   and from Chicago to Miami is 3 h 15 min.
   How much longer does it take to fly to Miami than to
   Minneapolis?

3. A bookshop is open from 9:30 a.m. to 5:00 p.m.
   How long is the bookshop open?

4. Molly went shopping at 10:20 a.m.
   She returned home 4 hours later.
   When did she return home?

5. Devi completed a jigsaw puzzle in 1 h 6 min.
   Lily completed the same jigsaw puzzle 10 minutes faster.
   How long did Lily take to complete the jigsaw puzzle?

6. Larry and his family went to the park for a picnic.
   They left home at 8:30 a.m. and arrived at the park at
   9:15 a.m.
   How long did the journey take?

7. Mr. Coles took 8 h 45 min to drive from Los Angeles to
   San Francisco.
   He arrived there at 2:15 p.m.
   What time did he leave Los Angeles?

REVIEW E

1. What time is it?
   (a) 8 h 55 min after 12:00 noon
   (b) 1 h 30 min after 12:00 midnight

2. Write the missing numerator or denominator.

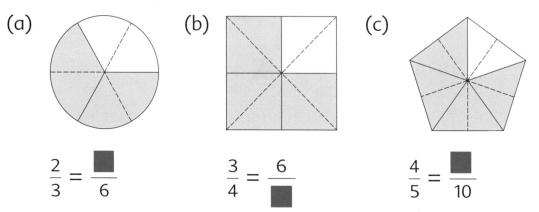

   (a) $\frac{2}{3} = \frac{\blacksquare}{6}$

   (b) $\frac{3}{4} = \frac{6}{\blacksquare}$

   (c) $\frac{4}{5} = \frac{\blacksquare}{10}$

3. This graph shows the number of vehicles in a parking lot.

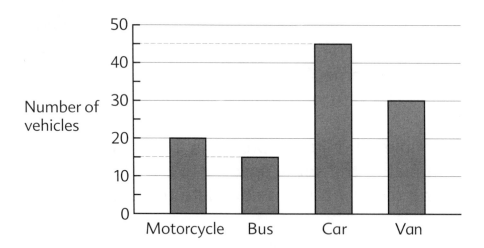

   (a) How many more vans than buses were there?
   (b) If there were 90 spaces for cars and vans, how many of them were **not** occupied?

4. Mr. Chen stayed in Japan for 19 months.
   Mr. Lee stayed there for 2 years 4 months.
   Who stayed longer?
   How many months longer?

5. A machine can fill 140 jars of jam in 10 minutes.
   How many jars can it fill in 1 minute?

6. An art lesson started at 5:40 p.m.
   It lasted 45 minutes.
   When did the lesson end?

7. A tank can hold 10 times as much water as a bucket.
   The capacity of the tank is 60 liters.
   What is the capacity of the bucket?

8. A piece of ribbon 1 m long is cut into two pieces.

   One piece is $\frac{5}{8}$ m long.

   What is the length of the other piece?

9. Adam bought a toothbrush and a tube of toothpaste.
   A toothbrush cost $6.50.
   A tube of toothpaste cost $1.80.
   How much did he pay altogether?

10. Tyler had a box of mangoes.
    After giving 3 mangoes each to 16 children, he had
    20 mangoes left.
    How many mangoes were there in the box at first?

11. Juanita spent $4.80 on strings and $2.50 on beads to make
    a flowerpot hanger.
    How much did it cost her to make a flowerpot hanger?

12. Mrs. Chen bought 8 towels.
    She gave the cashier $50 and received $2 change.
    (a) How much did she pay for the towels?
    (b) What was the cost of 1 towel?

Geometry

1 **Angles**

Use two cards to form an **angle** like this:

Then make a bigger angle.
What is the biggest angle you can get?
Compare it with your friends'.

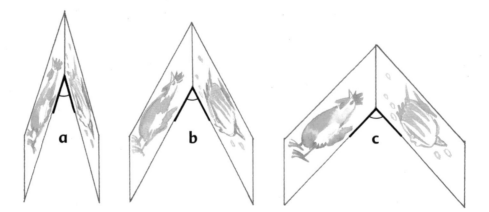

Which angle is the smallest?
Which angle is the biggest?

1. Here are some examples of angles.

Look for some more angles around you.

2. Two sides of a triangle make an angle.

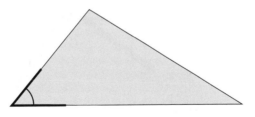

A triangle has ■ sides and ■ angles.

3. Here are some 4-sided figures.
   How many angles does each figure have?

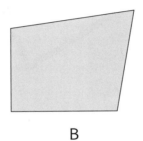

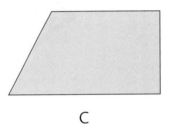

A   B   C

Workbook Exercise 45

Right Angles

Fold a piece of paper twice to make an angle like this:

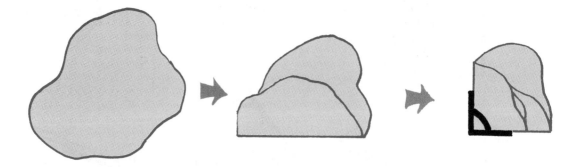

The angle you have made is a special one.
It is a **right angle**.
Use the right angle you have made to find out which of
the following angles are right angles.

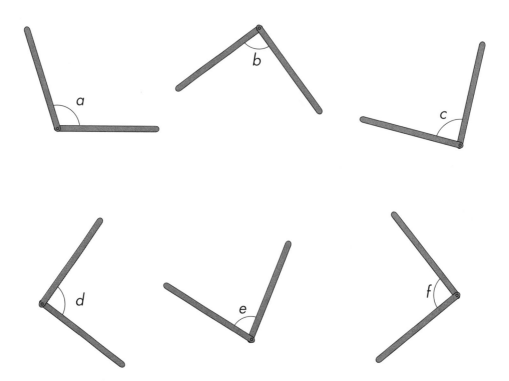

Use the right angle you have made to look for right angles
around you.

1. How many right angles can you find in
   (a) a square   (b) a rectangle?

2. Which one of these triangles has a right angle?
   Which one has an angle which is **greater than** a
   right angle?

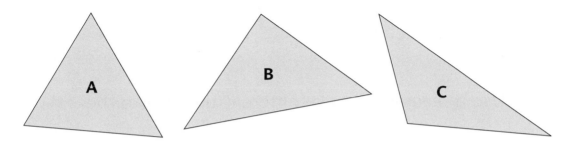

3. How many angles does each of these figures have?
   How many are right angles?

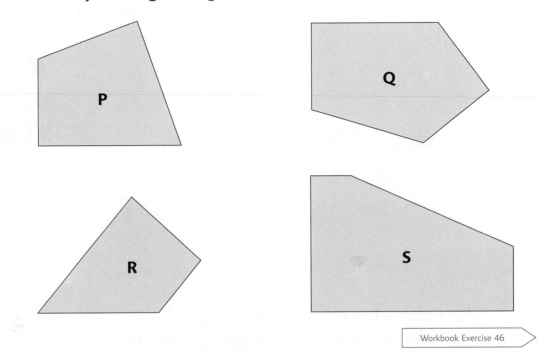

Workbook Exercise 46

95

Area and Perimeter

1 Area

Make 4 square cards and 4 half-square cards.

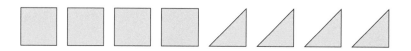

Use the cards to make different figures like these:

Each ▢ is 1 square unit.

Each ◿ is $\frac{1}{2}$ square unit.

The figures have the same **area**.
The area of each figure is ▇ square units.

1. What is the area of each of the following figures?

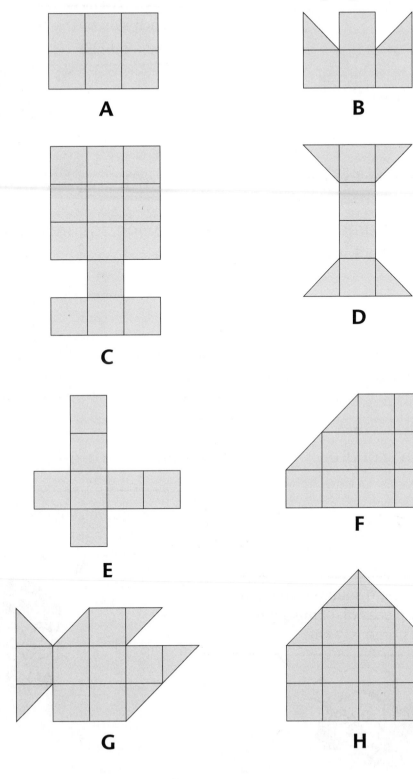

A

B

C

D

E

F

G

H

Which figure has the smallest area?
Which figure has the greatest area?

Workbook Exercise 47

2. This is a 1-cm square.

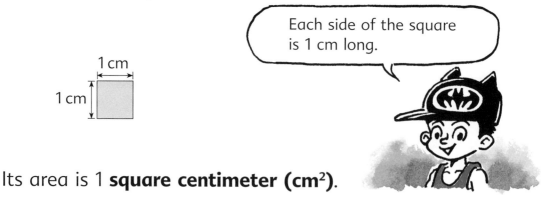

Each side of the square is 1 cm long.

Its area is 1 **square centimeter (cm²)**.

Give the area of each of the following squares in square centimeters.

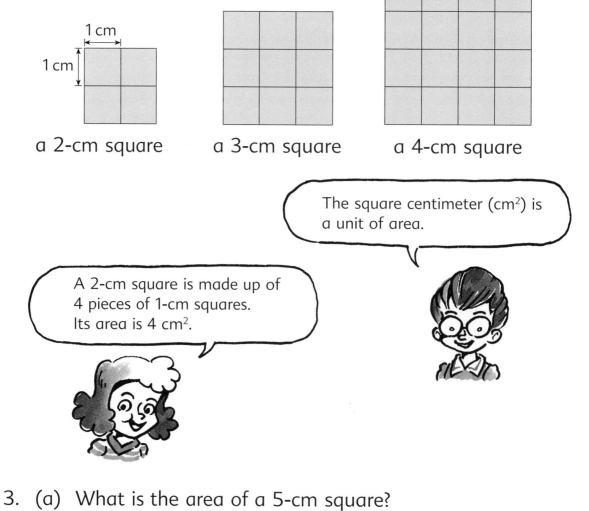

a 2-cm square   a 3-cm square   a 4-cm square

The square centimeter (cm²) is a unit of area.

A 2-cm square is made up of 4 pieces of 1-cm squares.
Its area is 4 cm².

3. (a) What is the area of a 5-cm square?
   (b) What is the area of a 10-cm square?

4. This figure is made up of 1-cm squares.
   Find its area.

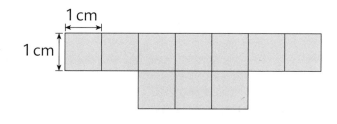

5. What is the area of each of the following figures?

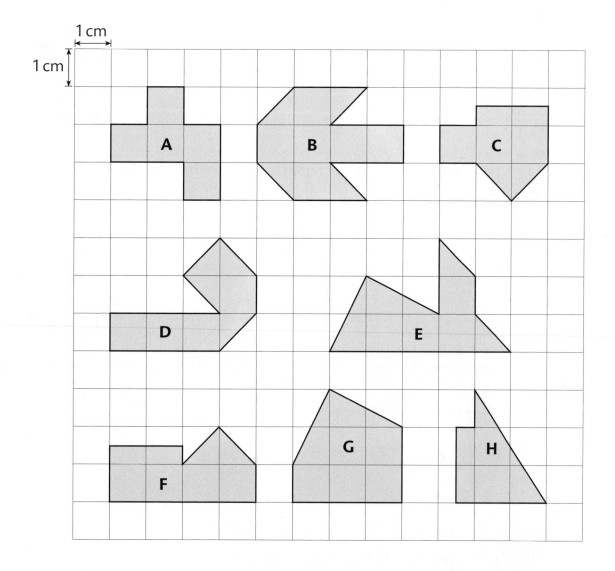

Workbook Exercise 48

6. Each side of this square is 1 m long.

Its area is 1 **square meter (m²)**.

The square meter (m²) is also a unit of area.

Give the area of each of the following figures in square meters.

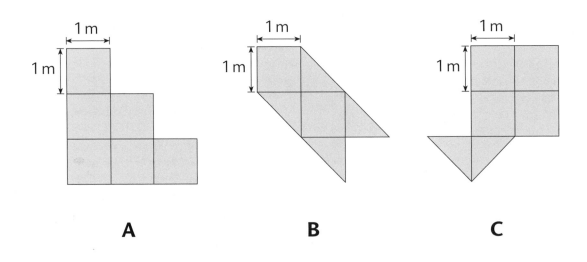

A

B

C

Which figure has the greatest area?
Which figure has the smallest area?

Workbook Exercise 49

2 Perimeter

Sumei used 3 pieces of wire of the same length to make the triangle, the square and the rectangle.

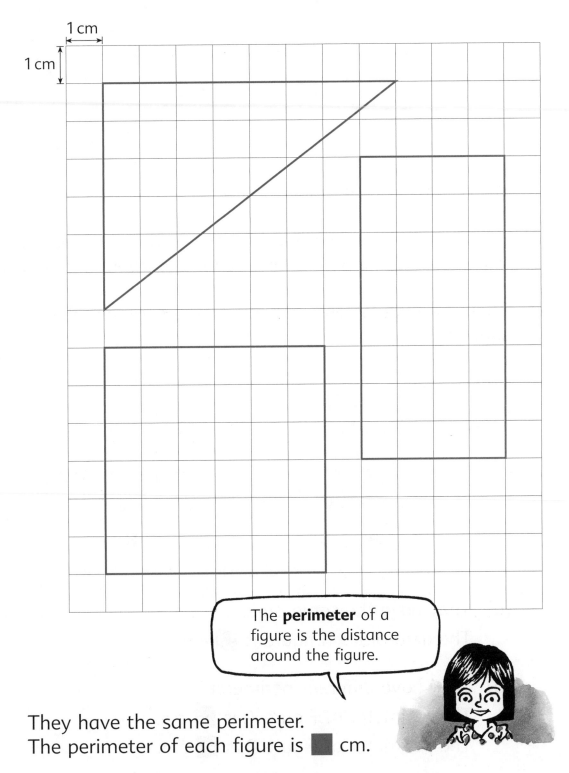

The **perimeter** of a figure is the distance around the figure.

They have the same perimeter.
The perimeter of each figure is ▮ cm.

1. Measure with thread, the perimeter of each of these figures.

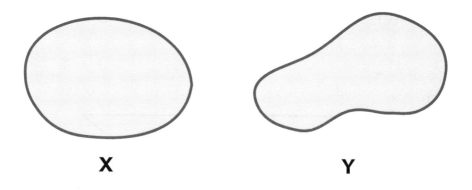

X   **Y**

Which figure has a longer perimeter?

2. (a) Measure the perimeter of your textbook in centimeters.
   (b) Measure the perimeter of your classroom in meters.

3. These two figures are made up of the same number of 1-cm squares.

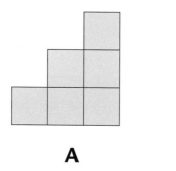 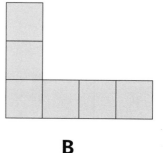

A   **B**

(a) They have the same area.
   The area of each figure is ■ cm².

(b) They have different perimeters.
   The perimeter of Figure A is ■ cm.
   The perimeter of Figure B is ■ cm.

4. These figures are made up of 1-cm squares.

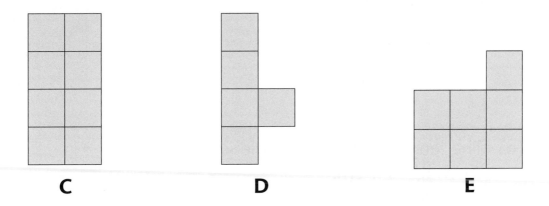

C   D   E

(a) Do they have the same area?
(b) Do they have the same perimeter?

5. These figures are made up of 1-cm squares.

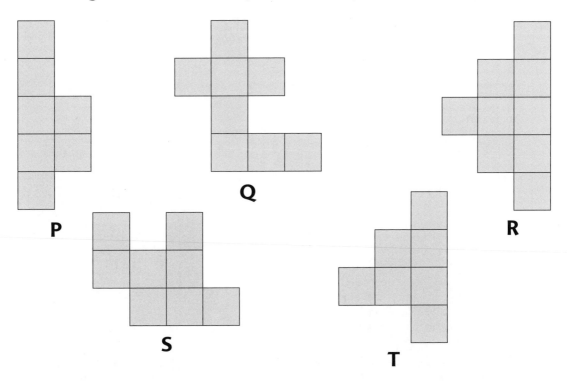

P   Q   R

S   T

(a) Which two figures have the same area but different perimeters?
(b) Which two figures have the same perimeter but different areas?
(c) Which two figures have the same area and perimeter?

6. (a) Each side of the square is 6 cm long.
Find its perimeter.

Perimeter = 6 + 6 + 6 + 6
= ■ cm

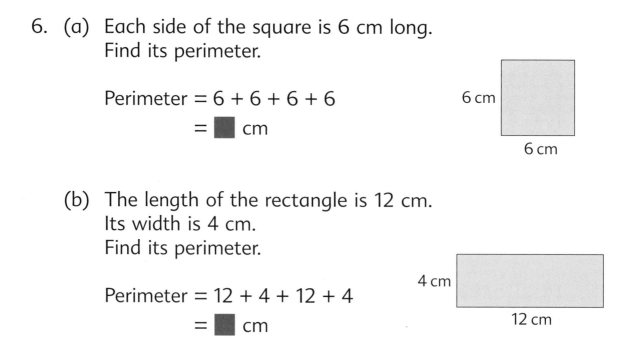

6 cm

6 cm

6 cm

(b) The length of the rectangle is 12 cm.
Its width is 4 cm.
Find its perimeter.

Perimeter = 12 + 4 + 12 + 4
= ■ cm

4 cm

12 cm

7. Find the perimeter of each of the following figures:

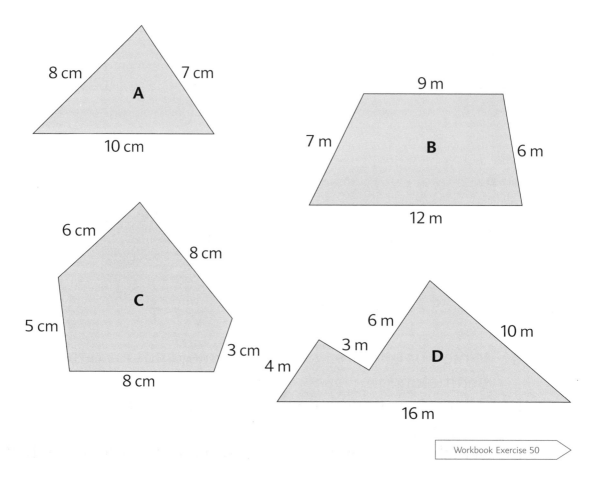

8 cm   7 cm
A
10 cm

9 m
7 m   B   6 m
12 m

6 cm
8 cm
C
5 cm
3 cm
8 cm
4 m

6 m
3 m
D
10 m
16 m

Workbook Exercise 50

③ Area of a Rectangle

Find the area of each of the following rectangles:

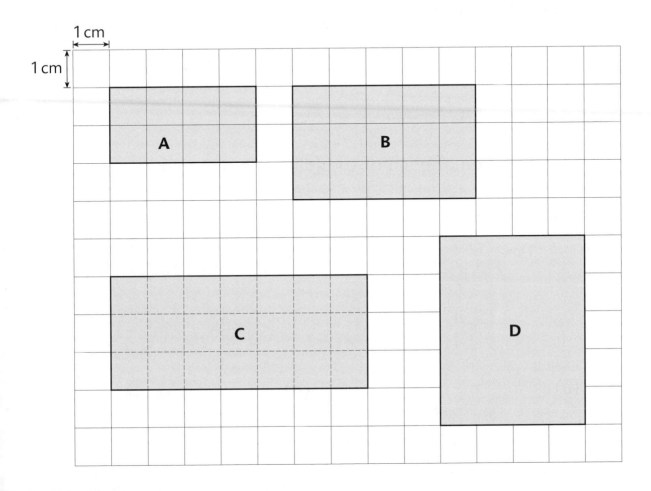

I count the square units covered by each rectangle to find its area.

I multiply the length and width of each rectangle to find its area.

Area of rectangle = Length × Width

1. Find the area of the rectangle.

Area of rectangle
= 5 × 4
= ▇ square units

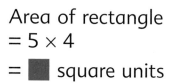

2. Find the area of each of the following rectangles:

(a)

(b)

(c)

4 cm

6 cm

(d)

9 m

3 m

(e)

20 cm

8 cm

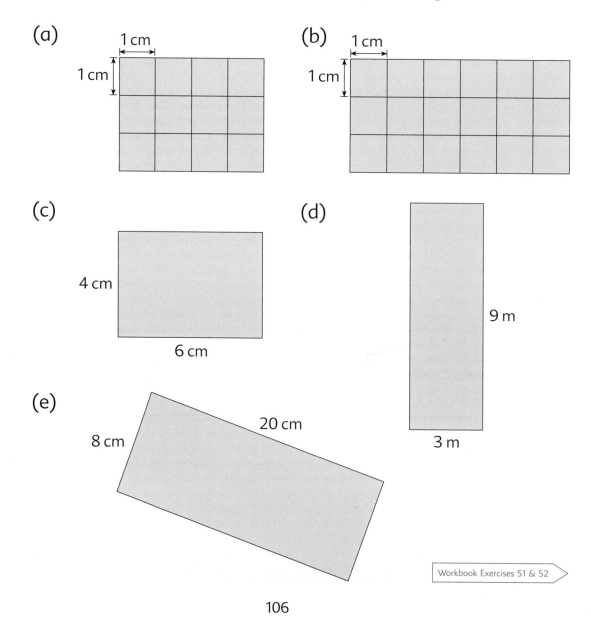

Workbook Exercises 51 & 52

PRACTICE 9A

1. Find the area and the perimeter of each of the following rectangles and squares:

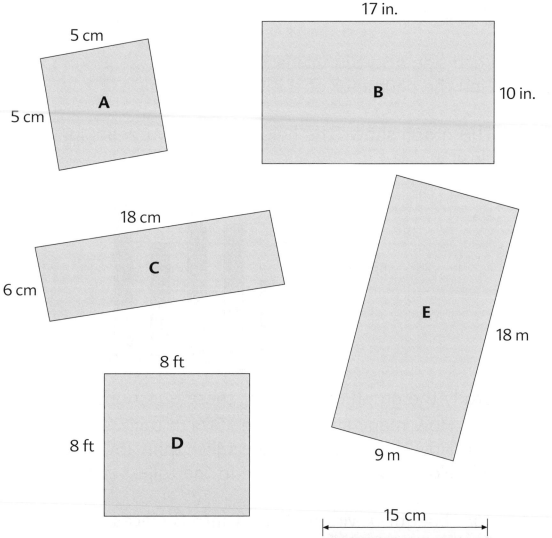

2. The length of the photograph is 15 cm.
   Its width is 10 cm.
   Find its area.

3. The length of a rectangular field is 85 m and its width is 10 m.
   Ian ran around the field once.
   How far did he run?

REVIEW F

1. Arrange the fractions in order, beginning with the smallest.

   (a) $\dfrac{3}{4}, \dfrac{1}{2}, \dfrac{5}{8}$

   (b) $\dfrac{1}{2}, \dfrac{3}{5}, \dfrac{3}{10}$

2. Each side of a triangle is 9 cm long.
   Find the perimeter of the triangle.

3. This graph shows the heights of the students in Mrs. Lee's class.

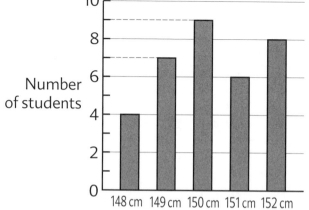

   Study the graph and answer these questions.
   (a) How many students are 150 cm tall?
   (b) How many students are taller than 150 cm?
   (c) How many students are in the tallest group?

4. Tom bought a wire and cut it into 8 pieces.
   Each piece of wire was 30 cm long.
   Find the length of the wire he bought.
   Give the answer in meters and centimeters.

5. 2500 people were at a show.
   1360 of them were men.
   240 were children.
   The rest were women.
   (a) How many women were there?
   (b) How many more adults than children were there?

6. Nicole made 286 cookies.
   She gave away 30 cookies and sold the rest at 8 for $1.
   How much money did she receive?

7. Each side of a square is 6 cm long.
   Find the area of the square.

8. Anna, Peter and John shared a pizza.

   Anna and Peter each received $\frac{1}{5}$ of the pizza.

   What fraction of the pizza did John receive?

9.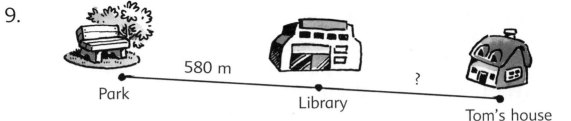

   580 m

   Park

   Library

   ?

   Tom's house

   Tom's house is 1 km from the park.
   How far is Tom's house from the library?

10. (a) What is the total weight of the
        3 pieces of butter and the
        bag of flour?
    (b) If each piece of butter weighs
        300 g, find the weight of the bag
        of flour in kilograms and grams.

11.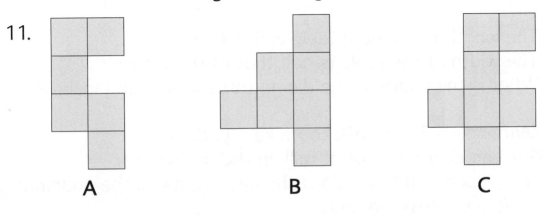

    A   B   C

    (a) Which two figures have the same area?
    (b) Which two figures have the same perimeter?

REVIEW G

1. Justin's school is one mile away from his home.
    His school is ▮ feet away from his home.

2. Write in inches.
    (a) 9 ft 6 in.   (b) 5 ft 10 in.   (c) 7 ft 7 in.

3. Add or subtract in compound units.
    (a) 5 lb 11 oz + 14 oz
    (b) 8 ft 8 in. + 5 ft 5 in.
    (c) 7 qt 1 pt + 11 qt 1 pt
    (d) 12 lb 2 oz − 6 lb 12 oz
    (e) 5 ft 1 in. − 2 ft 3 in.
    (f) 17 gal − 10 gal 1 qt

4. A box of candies weighs 1 lb 5 oz.
    A box of chocolate weighs 14 oz more than the candies.
    How much does the box of chocolate weigh?
    What is the total weight of the box of candies and the box
    of chocolate?

5. The total weight of 2 bags of sugar and one bag of flour is
    4 lb 2 oz. If the weight of each bag of sugar is 10 oz, find
    the weight of the bag of flour.

6. The length of a dining table is 5 ft 6 in.
    The width of the table is half that of the length.
    What is the width of the dining table in feet and inches?

7. Mary needs 3 c of milk to make a pudding.
    She can only find 1 pt of milk in the refrigerator.
    How much more milk does she need to make the pudding?
    (Give the answer in cups).

8. Eric bought 7 cans of paint.
   Each can contained 1 gal of paint.
   He used 3 gal 1 qt of paint to paint his room.
   How much paint was left?

9. Which container has the most water in it?

Container A	3 pt
Container B	1 qt
Container C	7 c
Container D	half gal

10. Find the perimeter of this figure.

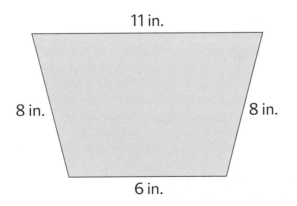

11 in.

8 in.   8 in.

6 in.

11. Mrs. Jackson paid $72 for some salmon.
    The cost of the salmon was $6 per pound.
    How many pounds of salmon did Mrs. Jackson buy?

12. A bag of beans weighing 2 lb 7 oz was divided equally into
    3 portions.
    What is the weight of each portion of beans in ounces?

13. A basket of oranges weighs 25 lb 3 oz.
    The empty basket weighs 1 lb 4 oz.
    What is the weight of the oranges?

14. A toy train is 19 in. long.
    Write the length in feet and inches.

15. Arrange the following lengths in order. Start with the longest.

String A	3 ft 1 in.
String B	29 in.
String C	2 ft 9 in.
String D	38 in.

16. Find the area of the following figures:

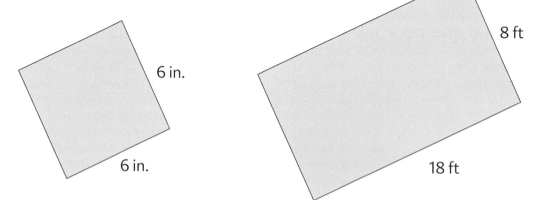

17. The capacity of a tank is 18 gal.
    How many quarts of water can it hold?

18. A piece of blue ribbon is 5 ft 4 in. long.
    A piece of yellow ribbon is 2 ft 8 in. shorter.
    What is the total length of the blue and yellow ribbons?

19. Tyrone weighs 20 lb 10 oz.
    Juan weighs 12 oz less than Tyrone.
    Sean weighs 1 lb 8 oz more than Tyrone.
    What is the total weight of the three boys?

20. Thomas ran round a rectangular field twice.
    The length of the field is 80 feet and its width is 50 feet.
    How far did Thomas run?